Undergraduate Texts in Mathematics

Readings in Mathematics

Editors
J. H. Ewing
F. W. Gehring
P. R. Halmos

Pierre Samuel

Projective Geometry

Translated from *Géométrie projective*, Presses Universitaires De France, by Silvio Levy

With 56 Illustrations

Springer-Verlag
New York Berlin Heidelberg
London Paris Tokyo

Pierre Samuel
Université de Paris-Sud (Orsay)
3, av. du Lycée-Lakanal
92350 Bourge-La-Reine
France

Silvio Levy *(Translator)*
Department of Mathematics
Princeton University
Princeton, NJ 08544
USA

Mathematics Subject Classification (1980): 51-01, 51A05

Projective Geometry by Pierre Samuel was originally published in French by Presses Universitaires de France, Paris, France, 1986; translated with permission.

Library of Congress Cataloging-in-Publication Data
Samuel, Pierre, 1921–
[Géometrie projective]
Projective geometry / Pierre Samuel ; translated by Silvio Levy.
p. cm.—(Undergraduate texts in mathematics. Readings in mathematics)
Translation of: Géometrie projective.
Bibliography: p.
Includes index.
ISBN 0-387-96752-4
1. Geometry, Projective. I. Title. II. Series.
QA471.S24 1988
516.5—dc19 88-12252

Camera-ready text prepared by the translator.
Printed and bound by R.R. Donnelley & Sons, Harrisonburg, Virginia.
Printed in the United States of America.

9 8 7 6 5 4 3 2 1

ISBN 0-387-96752-4 Springer-Verlag New York Berlin Heidelberg
ISBN 3-540-96752-4 Springer-Verlag Berlin Heidelberg New York

Introduction

When I was in high school, in the late thirties, I was fascinated by what was then, and had been since the late nineteenth century, called "modern geometry." This was in fact a type of algebra, so thoroughly known that actual calculations with coordinates had become almost unnecessary, and "pure-thought" reasonings sufficed. I was impressed by such statements as:

- If a line has at least three points in common with a conic, it belongs to the conic, which is the union of two lines.
- A conic having a double point is degenerate, as is a cubic having two and a quartic having four double points.
- The intersection of a quadric with a plane tangent to it is the union of two lines, and this defines two partitions of the quadric into families of lines.

And so it went with polarity, cubics, quartics, etc., up to the Villarceau circles on the torus. Just at the end we were introduced (without proof) to Bezout's theorem: "two plane curves of degree p and q have pq common points, real or imaginary, distinct or coincident, at finite distance or at infinity."

So it's not surprising that, a few years later, I specialized in algebraic geometry and, more exactly, in intersection theory. I did so in spite of (or perhaps because of) warnings from my friend Laurent Schwartz against the possibility of one's getting carried away and abusing principles like "a one-to-one algebraic correspondence (between lines, conics, and so on; sometimes even the adjective 'algebraic' is omitted) is a projective transformation".

Indeed, he would point out, let's take on the plane two complex conjugate lines D and D'; they intersect at a real point P. Every real line intersects

D at a point M and D' at its conjugate M', so this establishes between D and D' a one-to-one algebraic correspondence. Since the point P is in correspondence with itself, a well-known theorem (see section 2.5) says that all lines joining pairs of homologous points M, M' contain a fixed point. We have proved that all real lines of the plane pass through the same point!

This warning, plus the influence of my professor Claude Chevalley, who stressed the importance of finite fields and fields of characteristic p, made me pay particular attention to the choice of the field of scalars. The fallacy in the preceding argument, for example, is that the correspondence between M and M' is $\mathbf{R}$-rational, but not $\mathbf{C}$-rational. Furthermore, as I quickly found out, in characteristic $p \neq 0$ bijectivity is not a sufficient condition for a rational map to be a projective transformation. But, once the terminology has been made precise (and this involves a certain amount of algebra), all the beauty of the old "modern geometry" subsists. In any case, recent developments in algebraic geometry have shown that the old and the new blend quite harmoniously if one but sets up the necessary algebraic apparatus, whether the occasion be correspondences on a curve in characteristic $p \neq 0$ (André Weil), the proof of the Mordell conjecture (Gerd Faltings) or the formalization of old results of Halphen's on twisted curves (Gruson-Peskine).

* * *

On a more elementary and didactic note: I often had the occasion to use profitably what I had learned in school, sometimes spiced up with a bit of algebra, in preparing lessons for the candidates to the *Aggrégation* on the topics of projective geometry, conics and quadrics, and Möbius transformations. I had been considering the task of systematizing those somewhat scattered notes, made of bits and pieces, when my friend Francis Hirsch presented me with a good opportunity to do so by asking me to teach at the ENSET a course in algebra and geometry, primarily designed to expound the parts of the program that the prospective teachers knew less well, while at the same time enlarging their horizons. And the parts of the program that caused them most trouble were exactly those where algebra was applied to geometric situations that they were not very familiar with: projective, and sometimes affine, geometry; projective transformations, cross-ratios, conics, quadrics, and other curves or surfaces. The former situation had been reversed (following Jean Dieudonné's "down with Euclid"), and it seemed that geometry was no longer taught in high school (even to those preparing for the Ecole Polytechnique), or even in college. Thus I tried to share with those students the fun I had had in school, and to systematize the lectures I had been giving.

The present book is the result of this effort, and I very much enjoyed writing it. I decided to take the book a bit further than I had taken the course at the ENSET, where I had lots of things to teach that had nothing to do with projective geometry.

* * *

The prerequisites for reading this book are fairly limited: a good command of linear algebra, some knowledge of quadratic forms and field extensions, and a few notions about the irreducibility of polynomials and factorization in rings of polynomials.

While I tried to maintain the exposition on an elementary level, I decided not to limit it to **R** and **C**. Except for characteristic two, where the classification of conics and duality with respect to conics requires special care, there is no extra effort involved in studying characteristic p, and the results obtained can sometimes be unexpected (see section 3.4, for example) or amusing.

The figures, drawn over **R** (one hopes), are there just to help the reader grasp the relationships among points, lines and curves that appear in statements, and the reasonings which, in general, hold over (more or less) arbitrary fields. To simplify the drawing, all conics appear as circles; this is not a problem in general because we are discussing projective properties and every conic can be mapped onto a circle by a projective transformation.

Contents

CHAPTER 1

Projective Spaces

1.1. Projective Spaces and Projective Bases

Consider, in the plane, two non-parallel lines D and D', and a point P not contained in either line. To each point M of D associate the point $M' = p(M)$ where the line PM intersects D':

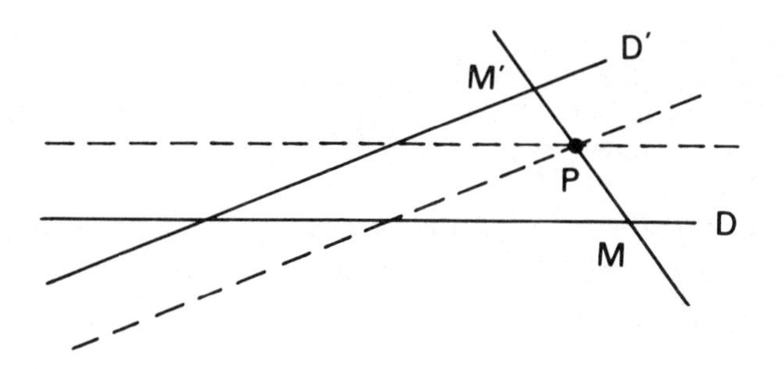

Notice that $p(M)$ is not defined when PM is parallel to D'; on the other hand, the point A' where D' intersects the parallel to D containing P is not in the image of p. There's something "missing" in D and D'; the right thing to work with seems to be the set of *projecting lines*, or lines containing the center P of projection. This motivates the following definition:

Definition 1. Given a vector space E over a field K, the *projective space* associated with E is the set $\mathbf{P}(E)$ of (vector) lines in E.

In this book the term "field" will include skew fields as well as commutative ones, except where we indicate otherwise (see index for a list of such sections). If K is a skew field we assume for concreteness that E is a left vector space over K.

One can also see $\mathbf{P}(E)$ as the quotient of the set $E\setminus 0$ of non-zero vectors modulo the equivalence relation "$x \sim y$ if and only if $y = ax$ for some $a \in K$" (naturally, $a \neq 0$). Thus we have a canonical map $p : E\setminus 0 \to \mathbf{P}(E)$ that associates to each vector x the vector line Kx it spans.

Definition 2. The *dimension* of $\mathbf{P}(E)$ is the integer $\dim E - 1$, which we denote by $\dim \mathbf{P}(E)$.

The projective space $\mathbf{P}(K^{n+1})$ is denoted by $\mathbf{P}_n(K)$; its dimension is n. Projective spaces of dimension one and two are called projective lines and planes, respectively.

Notice that $\mathbf{P}(0)$ is empty; by definition 2, its dimension is -1. A zero-dimensional projective space reduces to a point.

Definition 3. A *(projective) linear subvariety*, or *linear subspace*, of $\mathbf{P}(E)$ is the image $L = p(V)$ of a vector subspace V of E.

This definition embodies an abuse of notation: to be precise we should write $L = p(V \setminus 0)$.

Notice that a projective linear space $L = p(V)$ is the projective space $\mathbf{P}(V)$ associated with V.

An intersection of projective linear spaces is a (possibly empty) projective linear space. Given a subset $A \subset \mathbf{P}(E)$, there exists a smallest projective linear space containing A; we call it the projective linear space *generated* by A, and denote it (for the time being) by $v(A)$. It corresponds to the vector subspace spanned by $p^{-1}(A)$.

Theorem 1. *If L and L' are projective linear spaces in $\mathbf{P}(E)$, the following dimension formula holds:*

$$\dim L + \dim L' = \dim(L \cap L') + \dim\big(v(L \cup L')\big).$$

Proof. This is a direct translation, via definition 2, of the well-known formula for vector subspaces:

$$\dim V + \dim V' = \dim(V \cap V') + \dim(V + V'). \qquad \square$$

Corollary. *If $\dim L + \dim L' \geq \dim \mathbf{P}(E)$, the intersection $L \cap L'$ is non-empty.*

Proof. In fact, theorem 1 gives $\dim(L \cap L') \geq 0$, and the only empty projective linear space has dimension -1. $\square$

We say that a subset $A \subset \mathbf{P}(E)$ is *projectively independent* if it is the image under p of a linearly independent subset of E.

Homogeneous coordinates

We assume from now on that E is finite-dimensional.

Given a basis $(e_0, e_1, \ldots, e_n)$ for E, we can associate to each point $A \in \mathbf{P}(E)$ certain $(n+1)$-tuples of elements of K, namely, the coordinates of the vectors $x \in E$ such that $A = p(x)$. By definition, these $(n+1)$-tuples are all non-zero (that is, they have at least one non-zero component) and proportional to one another: if $(x_0, x_1, \ldots, x_n)$ is one $(n+1)$-tuple, all others will be of the form $(ax_0, ax_1, \ldots, ax_n)$, where $a \in K$ is non-zero. The set of such $(n+1)$-tuples is called the *homogeneous class* of $A \in \mathbf{P}(E)$, and each representative of this class is a set of *homogeneous coordinates* for A. The mapping thus defined from $\mathbf{P}(E)$ into the set of projective classes is called a *projective coordinate system*.

Projective coordinate systems can be characterized intrinsically in terms of $\mathbf{P}(E)$:

Theorem 2. *Let K be commutative.*

(a) *A projective coordinate system on an n-dimensional projective space $\mathbf{P}(E)$ is uniquely determined by the $n+2$ points with homogeneous coordinates* $(1,0,\ldots,0)$, $(0,1,\ldots,0)$, $\ldots$, $(0,0,\ldots,1)$, $(1,1,\ldots,1)$. *Any $n+1$ of these points form a projectively independent set.*

(b) *Conversely, for each $(n+2)$-tuple $(P_0, P_1, \ldots, P_{n+1})$ of points in $\mathbf{P}(E)$ all of whose $(n+1)$-subtuples are projectively independent, there exists a projective coordinate system assigning the coordinates* $(1,0,\ldots,0)$ *to* P_0, $\ldots$, $(0,0,\ldots,1)$ *to P_n and* $(1,1,\ldots,1)$ *to P_{n+1}.*

Proof. The $n+1$ points $P_0, \ldots, P_n$ are not enough to determine the basis of E from which the projetive coordinate system derives, because if $(e_0, e_1, \ldots, e_n)$ is such a basis, so is $(a_0e_0, a_1e_1, \ldots, a_ne_n)$ for any non-zero $a_0, \ldots, a_n \in K$. But if both bases assign to P_{n+1} the homogeneous coordinates $(1, \ldots, 1)$ we see that P_{n+1} is the image of both $e_0 + e_1 + \cdots + e_n$ and $a_0e_0 + a_1e_1 + \cdots + a_ne_n$, which implies that all the a_j are equal to the same non-zero scalar a. Thus the two bases are proportional, $(e_0, e_1, \ldots, e_n)$ and $(ae_0, ae_1, \ldots, ae_n)$.

Now if $M \in \mathbf{P}(E)$ comes from a point $x \in E$ whose coordinates in the first basis are $(x_0, x_1, \ldots, x_n)$, the coordinates of x in the second basis will be $(x_0a^{-1}, x_1a^{-1}, \ldots, x_na^{-1})$: the two sets of coordinates are proportional (that is, left-proportional) because K is commutative.

To prove part (b), lift $P_0, \ldots, P_n$ to any basis $(e_0, \ldots, e_n)$ of E, and consider the coordinates $(b_0, \ldots, b_n)$ of a vector $u \in p^{-1}(P_{n+1})$ in this basis. All the b_j are different from zero, so we just change our basis to $(b_0e_0, \ldots, b_ne_n)$. □

Part (b), and the last assertion in part (a), hold even if K is a skew field.

Definition 4. An $(n+2)$-tuple $(P_0, P_1, \ldots, P_{n+1})$ of points in $\mathbf{P}(E)$ is called a *(projective) frame* (or *projective base*) of $\mathbf{P}(E)$ if, for some projective coordinate system, the homogeneous coordinates of $P_0, \ldots, P_n$ are $(1, 0, \ldots, 0)$, $(0, 1, \ldots, 0)$, $\ldots$, $(0, 0, \ldots, 1)$, respectively, and those of P_{n+1} are $(1, \ldots, 1)$. The points $P_0, \ldots, P_n$ are called the *vertices*, and P_{n+1} the *unit point*, of the frame.

Corollary. *An $(n+2)$-tuple $(P_0, P_1, \ldots, P_{n+1})$ of points in an n-dimensional projective space is a projective frame if and only if any $n+1$ points among them are projectively independent.* □

It amounts to the same to say that no $(n-1)$-dimensional projective linear space, or *hyperplane*, contains $n+1$ of these points.

This corollary holds even if K is a skew field.

Examples

(1) A frame for a projective line is formed by any three distinct points ("pairwise distinct", as purists would have it).

(2) In a projective plane, a frame is formed by four points, three of which form a non-degenerate triangle and the fourth of which does not belong to any of the sides of the triangle. In this way no three points are collinear.

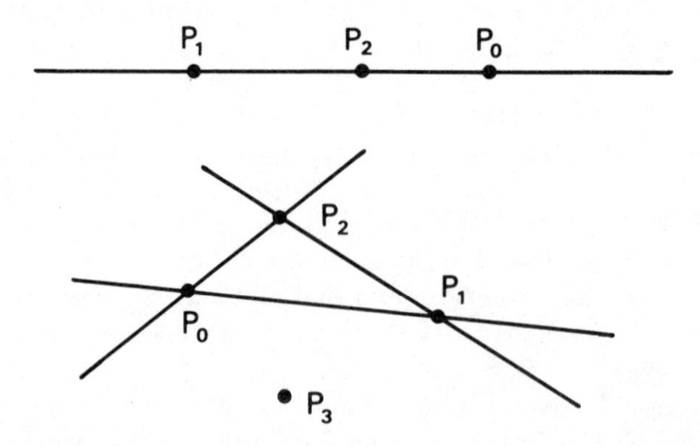

Homogeneous coordinates can be used to write equations for projective linear spaces of $\mathbf{P}(E)$. Namely, given a basis $(e_0, \ldots, e_n)$ for the vector space E, a hyperplane H has equation

(1) $\quad x_0 b_0 + x_1 b_1 + \cdots + x_n b_n = 0, \qquad$ with $b_j \in K$ not all zero,

which expresses the condition that the point with coordinates $(x_0, \ldots, x_n)$ be on H. The same equation (1) also expresses the condition that a point of $\mathbf{P}(E)$ with homogeneous coordinates $(x_0, \ldots, x_n)$ lies in the projective hyperplane $p(H)$; notice that any other set $(ax_0, \ldots, ax_n)$ of homogeneous coordinates for this point also satisfies (1).

As an intersection of hyperplanes, a projective linear space is defined by a system of homogeneous equations of the form (1). More precisely, if a projective linear space L has *codimension* d, that is, if its dimension is $n - d$, the space can be defined by a system of d equations whose left-hand sides are linearly independent linear forms.

Notice that in (1), the coefficients are written to the right of the variables. In fact, if f is a linear form having H as its kernel, we have $f(x_0 e_0 + \cdots + x_n e_n) = x_0 f(e_0) + \cdots + x_n f(e_n) = 0$.

If K is commutative, one defines an *algebraic subset* of E to be any subset given by a system of polynomial equations

$$F_j(x_0, x_1, \ldots, x_n) = 0 \qquad \text{for } j = 1, \ldots, q, \tag{2}$$

in some fixed basis of E. A change of basis alters these equations, but not their property of being polynomial, nor their degrees.

In translating this to the projective case, it's best to assume that the polynomials F_j are homogeneous. Then a system of equations of the form (2), if satisfied by one set of homogeneous coordinates of a point of $\mathbf{P}(E)$ in a given projective frame, is satisfied by the whole homogeneous class of the point. The equations are said to define an *algebraic subset* of $\mathbf{P}(E)$.

Cardinality over finite fields

Let K be the field $\mathbf{F}_q$ with q elements. If $\mathbf{P}(E)$ has dimension n, its characterization as a quotient space of $E \setminus 0$ immediately shows that

$$\#\mathbf{P}(E) = \frac{q^{n+1} - 1}{q - 1} = q^n + q^{n-1} + \cdots + q + 1. \tag{3}$$

Thus a projective line over $\mathbf{F}_q$ has $q + 1$ points (at least three, since $q \geq 2$), and a projective plane $q^2 + q + 1$ points.

The number of bases of E is $(q^{n+1} - 1)(q^{n+1} - q) \cdots (q^{n+1} - q^n)$, since we can start by choosing any non-zero vector, then any vector not proportional to the first, and so on. Since a projective frame is determined, up to a non-zero scalar factor, by a basis of E (theorem 2), we conclude that the number of frames of $\mathbf{P}(E)$ is

$$(q^{n+1} - 1)(q^{n+1} - q) \cdots (q^{n+1} - q^{n-1})q^n. \tag{4}$$

For lines and planes, respectively, the number of frames is $q(q^2 - 1) = q(q-1)(q+1)$ and $q^2(q^3 - 1)(q^3 - q) = q^3(q-1)^2(q+1)(q^2 + q + 1)$.

$\mathbf{P}(E)$ has as many d-dimensional projective linear spaces as E has $(d+1)$-dimensional vector subspaces. The number of such subspaces is the number of sets of $d+1$ linearly independent vectors in E, divided by the number of such sets as span the same subspace. This shows that the number of d-dimensional projective linear spaces of $\mathbf{P}(E)$ is

$$(5) \qquad \frac{(q^{n+1}-1)(q^{n+1}-q)\cdots(q^{n+1}-q^d)}{(q^{d+1}-1)(q^{d+1}-q)\cdots(q^{d+1}-q^d)}.$$

For q large this number is asymptotically $q^{(d+1)(n-d)}$.

In particular, the number of lines in $\mathbf{P}(E)$ is

$$\frac{(q^{n+1}-1)(q^n-1)}{(q-1)^2(q+1)}.$$

1.2. Projective Transformations and the Projective Group

Let u be a linear map from a vector space E into a vector space F. Since u preserves vector lines, it defines a map between the quotient spaces $\mathbf{P}(E)$ into $\mathbf{P}(F)$, as long as non-zero vectors are mapped into non-zero vectors, that is, u is one-to-one. The map $\mathbf{P}(u) : \mathbf{P}(E) \to \mathbf{P}(F)$ thus obtained is called a *projective map*, and a *projective transformation* if it is bijective, that is, if $\dim \mathbf{P}(E) = \dim \mathbf{P}(F)$. Projective transformations are sometimes called homographies.

When u is not one-to-one we obtain a map defined on the complement of $p(\ker(u))$.

Given another one-to-one linear map v from F into a third vector space G, we can write

$$(6) \qquad \mathbf{P}(v \circ u) = \mathbf{P}(v) \circ \mathbf{P}(u);$$

we also clearly have $\mathbf{P}(\mathrm{Id}_E) = \mathrm{Id}_{\mathbf{P}(E)}$.

Theorem 3. *Let E and F be vector spaces, with $\dim E \geq 2$, and let $Z = \{a \in K \mid ab = ba$ for all $b \in K\}$ be the center of K. Two one-to-one linear maps u and u' from E into F satisfy $\mathbf{P}(u) = \mathbf{P}(u')$ if and only if there exists a scalar $a \in Z$ such that $u'(x) = au(x)$ for every $x \in E$.*

The cases $\dim E = 0, 1$ are left to the reader.

Proof. The condition is obviously sufficient, since $a \in Z$ implies that $u \mapsto au$ is linear. Conversely, if $\mathbf{P}(u) = \mathbf{P}(u')$, there exists, for every non-zero $x \in E$, some scalar $a(x)$ such that $u'(x) = a(x)u(x)$. Taking x and y linearly independent and expressing $u'(x+y)$ in two different ways we find $a(x) = a(x+y) = a(y)$. This implies that $a(x) = a(y)$ for every x and y, since we can find z proportional to neither x nor y. There remains to show that $a = a(x)$ is central.

For any $b \in K$ and non-zero $x \in E$, we have $u'(bx) = au(bx) = abu(x)$ and $u'(bx) = bu'(x) = bau(x)$. Since $u(x) \neq 0$, this implies that $ab = ba$, showing that $a \in Z$. □

Corollary. *A one-to-one linear map of a vector space that transforms each vector into a multiple of itself is of the form $u \to au$, where a is a central scalar.* □

For K commutative this condition can also be stated in terms of eigenspaces. A map of the form $u \to au$ is called a *homothety*.

It follows from (6) that the projective transformations of $\mathbf{P}(E)$ into itself form a group, called the *projective group* of $\mathbf{P}(E)$ and denoted by $\mathrm{PGL}(E)$. Theorem 3 can be rephrased to say that if $\dim \mathbf{P}(E) \geq 2$ we have $\mathrm{PGL}(E) = \mathrm{GL}(E)/Z^*$, where $\mathrm{GL}(E)$ is the linear group of E and Z the center of K.

Notice that the fixed points of a projective transformation $\mathbf{P}(u)$ are the images of the (non-zero) eigenvectors of u.

Assume K commutative and fix a projective frame for $\mathbf{P}(E)$ (or, equivalently, fix a basis of E up to a scalar factor). A projective transformation of $\mathbf{P}(E)$ can be expressed in this basis by a class of proportional non-singular matrices, whose entries b_{ij} are defined by the condition that

$$(8) \qquad a y_j = b_{j0}x_0 + b_{j1}x_1 + \cdots + b_{jn}x_n \qquad \text{for } j = 0, 1, \ldots, n,$$

where $a \in K^*$ is arbitrary, $(x_0, \ldots, x_n)$ are the homogeneous coordinates of an arbitrary point in E and $(y_0, \ldots, y_n)$ the homogeneous coordinates of its image.

Theorem 4. *Let $\mathbf{P}(E)$ and $\mathbf{P}(E')$ be projective spaces of same dimension n over a commutative field K, with projective frames $(P_0, \ldots, P_n, P_{n+1})$ and $(P'_0, \ldots, P'_n, P'_{n+1})$, respectively. There exists a unique projective transformation $h : \mathbf{P}(E) \to \mathbf{P}(E')$ such that $h(P_i) = P'_i$ for all $i = 0, 1, \ldots, n, n+1$.*

Proof. Lift $(P_0, \ldots, P_n)$ to a basis $(e_0, \ldots, e_n)$ of E such that $p(e_0 + \cdots + e_n) = P_{n+1}$ (theorem 2), and lift $(P'_0, \ldots, P'_n)$ to $(e'_0, \ldots, e'_n)$. If h exists and is of the form $h = \mathbf{P}(u)$, each $u(e_i)$, for $i = 0, \ldots, n$, must be of the form $a_i e'_i$, where a_i is a non-zero scalar. Since $h(P_{n+1}) = P'_{n+1}$, the vector $u(e_0 + \cdots + e_n)$ can be written $b(e'_0 + \cdots + e'_n)$. Thus all the a_i are equal to b; this determines u up to a multiplicative scalar, and $h = \mathbf{P}(u)$ uniquely (theorem 3). The existence of u is obvious: define u by $u(e_i) = e'_i$ for $i = 0, 1, \ldots, n$. □

If K is skew, uniqueness fails. For example, take a and b in K such that $ab \neq ba$. If u is the linear map that takes the canonical basis $(e, f) = ((1,0),(0,1))$ of K^2

into (ae, af), it is easily checked that $h = \mathbf{P}(u)$ leaves invariant the points of the "canonical" frame of $\mathbf{P}(K^2)$ (those with homogeneous coordinates $(1,0)$, $(0,1)$ and $(1,1)$). But $\mathbf{P}(u)$ cannot be the identity, otherwise $u(e+bf) = ae + baf$ would be of the form $c(e+bf)$, which would imply $c = a$ and $ab = ba$.

Remark. Theorem 4 shows that, for K commutative, the number of elements of $\mathrm{PGL}(E)$ is equal to the number of frames of $\mathbf{P}(E)$. In particular, if K is finite, this number is given by formula (4).

1.3. Projective and Affine Spaces

Recap on affine spaces

Recall that an *affine space* is a set E on which the additive group of a vector space (denoted by $v(E)$ or $\vec{E}$) acts simply transitively. One often says that the elements of E are *points* and those of $v(E)$ are vectors, or translations of E; and $v(E)$ itself is called the vector space *underlying* E. The image of a point a under the translation t is generally denoted by $t+a$, whence the formulas

$$s + (t + a) = (s + t) + a,$$
$$0 + a = a,$$

which simply translate the fact that a group is acting on a set. The unique translation that takes a point a into a point b is denoted by $b - a$, or $\overrightarrow{ab}$. In this notation we have *Chasles's formula*

$$(c - b) + (b - a) = c - a. \tag{9}$$

The commutativity of the group $v(E)$ is equivalent to the *parallelogram rule*

$$b - a = b' - a' \quad \text{if and only if} \quad a - a' = b - b'. \tag{10}$$

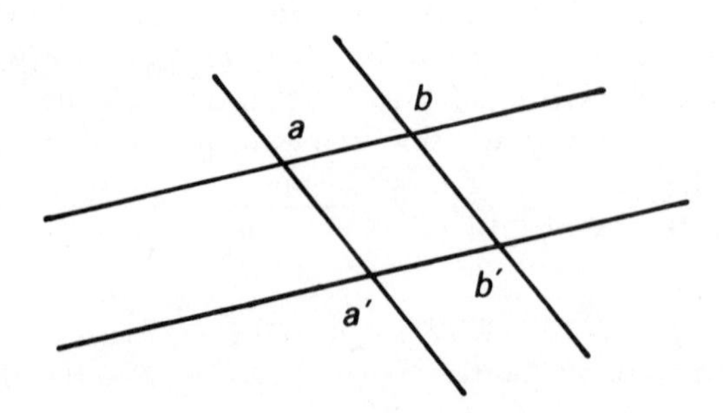

The choice of a point $a \in E$ allows one to identify E with its underlying vector space $v(E)$: each point $m \in E$ gets associated with the vector $m - a \in v(E)$. Although this choice of an origin, or *vectorialization*, is by no means intrinsic, one often performs it in order to make calculations easier.

An *affine frame* of E is made up of a point a_0 and a basis $(e_1, \ldots, e_n)$ of $v(E)$. The coordinates of a point m in this frame are those of the vector $m - a_0$ in the given basis. It amounts to the same to give the $n+1$ points a_0, $a_1 = e_1 + a_0$, ..., $a_n = e_n + a_0$.

An *affine linear subspace* (or *subvariety*) L is a subset of E that is either empty or of the form $L = V + a$, where V is a vector subspace of $v(E)$ and a is a point in E. Since $V + a = V' + a'$ if and only if $V = V'$ and $a' - a \in V$, the vector subspace V is uniquely determined by L; it is called the *direction* of L. Two affine subspaces are called *parallel* if they have the same direction. When we choose an origin for E, affine subspaces (other than the empty one) are simply translates of vector subspaces of $v(E)$. Every intersection of affine subspaces is one, so we have the notion of the affine subspace generated by a subset $A \subset E$.

Given points $m_1, \ldots, m_q \in E$ and scalars $a_1, \ldots, a_q \in K$ such that $a_1 + \cdots + a_n = 1$, we define the *barycenter* of the m_i with weights a_i as the unique point g such that

$$g - p = a_1(m_1 - p) + \cdots + a_q(m_q - p)$$

for every $p \in E$.

The operation of taking barycenters is "associative": a barycenter of barycenters of points m_i is a barycenter of the m_i. It can be shown that the set of barycenters of a set of points m_i is just the affine subspace generated by the m_i. In particular, if a subset $S \subset E$ is invariant under the operation of taking barycenters of sets of points, S is an affine subspace.

Remark. For $K \neq \mathbf{F}_2$ a subset invariant under the operation of taking barycenters of *two points* is an affine subspace. But over $\mathbf{F}_2$ the barycenters of two points are just the two points, so all subsets are invariant under this operation.

In an affine frame, with coordinates denoted by $(x_1, \ldots, x_n)$, an affine subspace is defined by a system of linear equations

$$x_1 a_{j1} + \cdots + x_n a_{jn} = b_j \qquad \text{for } j = 1, \ldots, q,$$

where the a_{ji} and the b_j are scalars.

One can choose the linear forms so that their left-hand sides are linearly independent; then the dimension of the affine subspace is $n - q$. (The dimension of an affine subspace is the dimension of its direction.)

This is only true about non-empty affine subspaces. The empty affine subspace can be defined by the equations $x_1 = 0$ and $x_1 = 1$, for example.

Example: the complement of a hyperplane in a projective space

Theorem 4. *Let P be an n-dimensional projective space and $H \subset P$ a hyperplane. Denote by T the set containing the identity and the projective transformations of P that leave invariant exactly those points of P that belong to H. Then T is a group isomorphic to the additive group of an n-dimensional vector space, and it acts simply transitively on $P \setminus H$.*

Proof. Write $P = \mathbf{P}(V_1)$ and $H = \mathbf{P}(H_1)$, where V_1 is a vector space over K and $H_1 \subset V_1$ a hyperplane. Choose $z \in V_1$ such that V_1 is the direct sum of H_1 and Kz, and consider $u \in \mathrm{GL}(V_1)$ such that $\mathbf{P}(u) \in T$ and $\mathbf{P}(u) \neq \mathrm{Id}$. Since $\mathbf{P}(u)$ leaves invariant all points in H, there exists $a \in K$ such that $u(x) = ax$ for every $x \in H_1$. On the other hand, we can write $u(z) = h + bz$, with $h \in H_1$ and $b \in K$.

A fixed point of $\mathbf{P}(u)$ comes from an eigenvector of u; if we write such a vector in the form $x + cz$, with $x \in H_1$ and $c \in K$, the condition is that $u(x+cz) = d(x+cz)$ for some $d \in K$, that is, that $ax+c(h+bz) = dx+dcz$. This is equivalent to the system

$$\text{(S)} \qquad \begin{aligned} (a-d)x + ch &= 0, \\ cb &= dc. \end{aligned}$$

Thus $\mathbf{P}(u)$ has a fixed point outside H if and only if there exists a solution (c, d, x) of (S) with $c \neq 0$. If $h = 0$, there exists such a solution with $c = 1$, $d = b$ and $x = 0$; the assumption $\mathbf{P}(u) \in T$ then requires $\mathbf{P}(u) = 1$, whence $a = b$. If $h \neq 0$, on the other hand, (S) and $c \neq 0$ together imply $d = cbc^{-1}$ and $(a - cbc^{-1})x = -ch$; this has a solution unless $a - cbc^{-1}$ vanishes for every non-zero c, that is, unless a and b are equal and belong to the center of K.

Thus if $\mathbf{P}(u) \in T$ we can normalize u so as to make $a = b = 1$; then u is the identity on H_1 and $u(z)$ is of the form $u(z) = h_u + z$, where $h_u \in H_1$ is uniquely determined by $\mathbf{P}(u)$. Furthermore, it's easy to see that $h_{u \circ v} = h_u + h_v$ if $\mathbf{P}(u), \mathbf{P}(v) \in T$. Hence the group structure on T. That T acts simply transitively on $P \setminus H$ follows from the fact that the translations h_u act simply transitively on $z + H_1$. □

Embeddings of an affine space in a projective space

Let E be an affine space over K, with underlying vector space $v(E)$, and $a \in E$ a point. The *projective closure* $\hat{E}$ of E is defined as

$$\hat{E} = \mathbf{P}\big(v(E) \times K\big). \tag{11}$$

We also define an injection $j_a : E \to \hat{E}$ by

$$j_a(m) = p(m - a, 1) \qquad \text{for } m \in E. \tag{12}$$

This injection is determined, up to a translation, by the choice of a. The image $j_a(E)$ is the complement of the hyperplane $\mathbf{P}\big(v(E) \times 0\big)$ in $\hat{E}$.

Let E and F be affine spaces over K. A map $f : E \to F$ is called an *affine map* if there exists a linear map $v(f) : v(E) \to v(F)$ such that $v(f)(m - m') = f(m) - f(m')$ for every $m, m' \in E$. If $f : E \to F$ is affine and one-to-one, it defines a projective map

$$\hat{f} = \mathbf{P}\big(v(f) \times \mathrm{Id}_K\big) : \hat{E} \to \hat{F}. \tag{13}$$

This map extends f in the sense that

$$\hat{f}\big(j_a(m)\big) = j_{f(a)}\big(f(m)\big) \qquad \text{for every } m \in E; \tag{14}$$

this is because $\hat{f}\big(j_a(m)\big)$ is the canonical image in $\hat{F}$ of $\big(f(m) - f(a), 1\big)$.

With the obvious notations, we have

$$\hat{u} \circ \hat{v} = \widehat{u \circ v}, \tag{15}$$

so there is a canonical injection from the affine group of E into the projective group of $\hat{E}$. Its image consists of the projective transformations that leave the *hyperplane at infinity* $\mathbf{P}\big(v(E) \times 0\big)$ globally invariant.

Affine coordinates and homogeneous coordinates

Thus every affine space E can be seen as the complement $P \setminus H$ of a hyperplane in a projective space. This hyperplane and all sets lying in it are said to be *at infinity*. Notice that two affine subspaces are parallel if and only if they have the same points at infinity (more rigorously, we should say that their projective completions have the same points at infinity, but we won't be so sticky).

Conversely, if we pick a hyperplane H in a projective space P and concentrate on the affine structure of $P \setminus H$, we'll often say that H is the hyperplane at infinity, or that H has been *sent to infinity*. We can then choose a projective coordinate system $(x_0, x_1, \ldots, x_n)$ on P in such a way that H is the hyperplane of equation $x_0 = 0$. If $m \in P$ is a point not on H, the n-tuple $(x_0^{-1}x_1, \ldots, x_0^{-1}x_n)$ gives the affine coordinates of m. If a projective linear space L of P which is not at infinity is given by the system of (homogeneous) equations

$$x_0 a_{j0} + x_1 a_{j1} + \cdots + x_n a_{jn} = 0 \qquad \text{for } j = 1, \ldots, q,$$

its intersection with $P \setminus H$ is the affine subspace defined by the (affine) equations $a_{j0} + y_1 a_{j1} + \cdots + y_n a_{jn} = 0$. If K is commutative and L is an algebraic subset of P, not contained in H, and defined by the homogeneous polynomial equations

$$F_j(x_0, \ldots, x_n) = 0 \qquad \text{for } j = 1, \ldots, q,$$

the intersection $L \cap (P \setminus H)$ is defined by the equations $F_j(1, y_1, \ldots, y_n) = 0$. In sum, to pass from projective to affine equations, just take $x_0 = 1$.

When L is at infinity, the system of affine equations obtained by this procedure is "impossible", that is, it has no solutions, even over the algebraic closure of K.

Conversely, let E be an affine space with a fixed affine frame, and denote by $(y_1, \ldots, y_n)$ the coordinates of a point $m \in E$. Embed E in $\hat{E} = P$ using the injection j_a associated with the origin a of the chosen frame. By (11) and (12), $(1, y_1, \ldots, y_n)$ is a set of homogeneous coordinates for $j_a(m)$, in the corresponding projective coordinate system of $K \times v(E)$. If L is an affine subspace of E defined by the equations

$$y_1 a_{j1} + \cdots + y_n a_{jn} = b_j \qquad \text{for } j = 1, \ldots, q,$$

the projective closure $\hat{L}$ of L is given by the equations $x_1 a_{j1} + \cdots + x_n a_{jn} = x_0 b_j$, and we have $L = (P \setminus H) \cap \hat{L}$, where H is the hyperplane at infinity.

Now assume that K is commutative and consider an algebraic subset A of the affine space E, defined by a single polynomial equation $F(y_1, \ldots, y_n) = 0$. Denote by d the total degree of F, and form the homogeneous polynomial F_h of degree d associated with F:

$$F_h(x_0, x_1, \ldots, x_n) = x_0^d F(x_1/x_0, \ldots, x_n/x_0). \tag{15}$$

The algebraic subset of P defined by the homogeneous equation $F_h(x_0, x_1, \ldots, x_n) = 0$ is called the *projective closure* of A, and is denoted by $\hat{A}$. Since $F_h(1, y_1, \ldots, y_n) = F(y_1, \ldots, y_n)$, the set A is the intersection of $\hat{A}$ with $P \setminus H$. The points of $\hat{A} \setminus A$ are called points at infinity of A; they make up an algebraic subset of H.

In this discussion we have limited ourselves to hypersurfaces, or algebraic sets defined by a single equation (hypersurfaces are called curves or surfaces if $n = 2$ or 3, respectively). The dimension of such objects, whether affine or projective, is $n - 1$, by any reasonable definition.

In treating lower-dimensional algebraic subsets of E, defined by several polynomial equations, it's not enough to homogenize the defining equations; one must also homogenize all the polynomials in the ideal generated by them.

For example, consider in $\mathbf{C}^3$ the circle C defined by the equations $x^2 + y^2 + z^2 - 1 = 0$ and $x^2 + y^2 + z^2 - 2x = 0$. The corresponding homogeneous equations are $x^2 + y^2 + z^2 - t^2 = 0$ and $x^2 + y^2 + z^2 - 2xt = 0$ (where the homogenizing variable is written t instead of x_0). The algebraic set defined by these two equations is the union of C with a curve at infinity, of equation $x^2 + y^2 + z^2 = t = 0$, which is called an umbilic. But the actual projective closure of C is smaller than that: it has only two points at infinity, where it intersects the umbilic. The reason is that the polynomial $2x - 1$, for example, is in the ideal generated by $x^2 + y^2 + z^2 - 1$ and $x^2 + y^2 + z^2 - 2x$, so by definition points in $\hat{C}$ must be zeros of the homogenized polynomial $2x - t = 0$. In this case we can get around the problem of extra points at infinity by replacing one of the two equations of spheres that define C by the equation $2x - 1 = 0$ of their radical plane. There are cases, however, where no such replacement is possible.

Given a projective space P and a system of projective coordinates for it, say $(x_0, x_1, \ldots, x_n)$, the hyperplanes H_i of equation $x_i = 0$, for $i = 0, \ldots, n$, have empty intersection, which means that P is the union of the $n + 1$

affine spaces $P \setminus H_i$. The affine coordinates in $P \setminus H_i$ of a point whose homogeneous coordinates are $(x_0, x_1, \ldots, x_n)$ are given by

$$(x_i^{-1}x_0, \ldots, x_i^{-1}x_{i-1}, x_i^{-1}x_{i+1}, \ldots, x_i^{-1}x_n).$$

Consider a point in $P \setminus (H_0 \cup H_i)$, and let its affine coordinates in $P \setminus H_0$ be $(y_1, \ldots, y_n)$; by assumption, $y_i \neq 0$. The homogeneous coordinates of this point are $(1, y_1, \ldots, y_n)$, so its affine coordinates in $P \setminus H_i$ are

$$(y_i^{-1}, y_i^{-1}y_1, \ldots, y_i^{-1}y_{i-1}, y_i^{-1}y_{i+1}, \ldots, y_i^{-1}y_n). \tag{16}$$

In practice we allow ourselves some abuses in notation. For example, if we start from the affine curve C defined by $x^3 + xy + 1 = 0$ and denote by z the homogenizing variable, the projective closure $\hat{C}$ of C is given by $x^3 - xyz + z^3 = 0$; in order to study the point $(x, y, z) = (0, 1, 0)$, the only point at infinity of the closure, we can make $y = 1$, obtaining the equation $x^3 + z^3 - xz = 0$ for the "affine piece" of $\hat{C}$ that lies in the affine space $y \neq 0$. Obviously the letters x, y, z don't have the same meaning in the three equations.

Simple and multiple points

Here we assume that K is commutative and infinite. Let V be an affine hypersurface with equation $F(y_1, \ldots, y_n) = 0$. We will write the polynomial F in the form

$$F(Y) = F_0 + F_1(Y) + \cdots + F_d(Y),$$

where $Y = (Y_1, \ldots, Y_n)$, F_j is homogeneous of degree j and $F_d \neq 0$. The integer d is called the *degree* of F. Let D be the affine line defined by $y_i = a_i + b_i t$, where $i = 1, \ldots, n$ and $t \in K$ is a parameter; D goes through the point $A = (a_1, \ldots, a_n)$. The parameter values at the intersections of D with V are the roots of the equation

$$F(a_1 + b_1 t, \ldots, a_n + b_n t) = 0. \tag{18}$$

This equation is identically satisfied if and only if V contains D, since we assumed K infinite; from now on we exclude this case. Otherwise (18) has degree at most d, so D has at most d distinct common points $P_1, \ldots, P_r$ with V. Let $t_1, \ldots, t_r$ be their parameters. The multiplicity m_j of the root t_j of (18) depends only on the point P_j; it remains the same if we change frames or if we change the parameter along D (by an affine transformation). This number m_j is called the *intersection multiplicity* of V and D at P_j.

If the point $A = (a_1, \ldots, a_n)$ is on V, (18) has a root at $t = 0$. This root is simple if and only if the coefficient of t in (18) is non-zero. By Taylor's formula, this coefficient is $F_1'(a)b_1 + \cdots + F_n'(a)b_n$, where $F_i'(a)$ is the i-th partial derivative of F at $(a_1, \ldots, a_n)$. If at least one partial derivative at A is non-zero, we say that A is a *simple point* of V. Then the root $t = 0$ is simple unless the vector $(b_1, \ldots, b_n)$ satisfies $F_1'(a)b_1 + \cdots + F_n'(a)b_n = 0$;

this condition amounts to saying that the tip $(y_1, \ldots, y_n)$ of the vector is on the hyperplane

$$F_1'(a)(y_1 - a_1) + \cdots + F_n'(a)(y_n - a_n) = 0, \tag{19}$$

called the *tangent hyperplane* to V at A. The lines of this hyperplane that go through A and whose intersection multiplicity with V at A is at least two are said to be *tangent* to V at A.

A point $A = (a_1, \ldots, a_n)$ of V such that $F_1'(a) = \cdots = F_n'(a) = 0$ is said to be *multiple* (or *singular*); such points form an algebraic subset of V, defined by $n+1$ equations. To study such a point more closely, we make it the origin; then $F_0 = F_1 = 0$ in (17). Let m be the smallest integer such that the homogeneous polynomial F_m is non-zero; m is called the *multiplicity* of A on V. Equation (18) becomes

$$F_m(b_1, \ldots, b_m)t^m + \cdots + F_d(b_1, \ldots, b_m)t^d = 0; \tag{20}$$

thus the intersection multiplicity of V and D at A is m unless D lies in the *tangent cone* of equation $F_m(y_1, \ldots, y_m) = 0$.

Assume now that all the roots of (18) are in K (for example, if K is algebraically closed). If (18) has maximal degree, namely d, we can say that D and V have d common points, where each point P_j is counted with its intersection multiplicity m_j. But if (18) has degree less than d, because its highest coefficient $F_d(b_1, \ldots, b_n)$ vanishes, it's no longer true that V and D have d common points. Where are the other points gone? To infinity, of course. Indeed, the relation $F_d(b_1, \ldots, b_n) = 0$ implies that the point at infinity of D is in V (or rather, in $\hat{V}$); we say then that the direction of D is an *asymptotic direction* of V.

Examples. The intersection of the plane curve $x^4 - y^4 - xy = 0$ with the line $y = bx$ is determined by the equation $(1 - b^4)x^4 - bx^2 = 0$. This equation has $x = 0$ as a double root, that is, the origin is a double point. The degree drops to 2 for $b = 1, -1, i, -i$, which are the slopes of the asymptotic directions of the curves.

The surface $x^2 + y^3 + z^5 = 0$ has only one multiple point in affine space, the origin: if the partial derivatives $2x$, $3y^2$, $5z^4$ all vanish we have $x = y = z = 0$ in characteristic $\neq 2, 3, 5$, and in characteristic 2, 3 or 5 two of the coordinates are zero, hence so is the third because of the equation of the surface. Its asymptotic directions, those along which the highest-degree term vanishes, are the directions contained in the plane $z = 0$. The intersection $z = t = 0$ of this plane with the plane at infinity is the part at infinity of the projective closure of the surface. All points in this intersection are singular (in the closure); to see this, one can observe that the degree of (18) drops by 2 in all directions such that $z = 0$, or else write the equation of the surface in the affine patches $x \neq 0$ and $y \neq 0$, say, and take partial derivatives (the equation in the patch $y \neq 0$, for example, is $t^2 + x^2t^3 + z^5 = 0$).

Thus we're led to consider, in a projective space P, the intersection of a hypersurface V of homogeneous equation $G(x_0, \ldots, x_n) = 0$ with a line D of parametric equation

$$x_i = c_i u + d_i v \qquad \text{for } i = 0, \ldots, n,$$

where the "parameter" (u, v) in K^2 is to be understood projectively, that is, $(u, v) \neq (0, 0)$, and two proportional pairs parametrize the same point. The intersection of V and D is governed by the equation

$$G(c_0 u + d_0 v, \ldots, c_n u + d_n v) = 0. \tag{21}$$

This is a homogeneous equation of degree $d = \deg G$ in u and v. Replacing K, if necessary, by an algebraic extension, we can write the left-hand side of (21) as a product of linear factors in (u, v), as follows: factor out u^k, for $k \geq 0$ maximal, then solve the equation obtained by making $u = 1$; each root e_j of this equation yields a factor $v - e_j u$ of (21). Counting each factor with its exponent, we obtain d solutions (u, v), each of which can be put in the form $(0, 1)$ or $(1, e_j)$. Thus we obtain exactly d points common to V and D. In the old literature this is expressed by saying that V and D have d common points, "real or imaginary" (replace K by its algebraic closure), "distinct or not" (count multiplicities), "at finite distance or at infinity" (replace the affine hypersurface by its closure).

Thus we see where the "disappearing" intersection points go when equation (18) drops from degree d to degree $d - k$. The idea is to take G above to be the homogeneous polynomial associated with F, and the c_i and d_i $(i = 1, \ldots, n)$ to describe the same line D whose affine representation is $y_i = a_i + b_i t$: writing $x_0 = u$ and $x_i = a_i u + b_i v$, for example, we get $c_0 = 1$, $d_0 = 0$, $c_i = a_i$, $d_i = b_i$. Then (21) becomes

$$G(u, a_1 u + b_1 v, \ldots, a_n u + b_n v) = 0. \tag{22}$$

Upon setting $v = tu$ this equation becomes $u^n G(1, a_1 + b_1 t, \ldots, a_n + b_n t) = 0$, which reduces to (18) if $u \neq 0$. The $d - k$ roots $t_j \in K$ of (18) account for the $d - k$ factors $v - t_j u$ in the left-hand side of (22); but there are also k factors u, corresponding to the point at infinity of D, counted k times. Notice that the intersection multiplicities are the same in the affine and the projective cases.

Finally, let's spell out the projective version of the notions of simple points and tangent hyperplanes. Assume that a point A, with homogeneous coordinates $(c_0, \ldots, c_n)$, lies on V, so that the coefficient of u^d in the left-hand side of (21) is zero. D and V intersect at A with multiplicity one if and only if v is a simple factor in the left-hand side of (21), if and only if the coefficient of $u^{d-1}v$ is non-zero. By Taylor's formula, this coefficient is $d_0 G_0'(c) + \cdots + d_n G_n'(c)$, where $G_i'(c)$ is the i-th partial derivative of G evaluated at $(c_0, \ldots, c_n)$.

Now A is a simple point of V if any line intersecting V at A does so with multiplicity one; by the previous paragraph, this happens if and only if at least one of the $G'_i(c)$ is non-zero. Thus the multiple points of V are defined by the $n+2$ equations $G(x) = G'_i(x) = 0$. If A is a simple point of V the *tangents* to V at A (that is, the lines whose intersection with V at A has multiplicity at least 2) are characterized by belonging to the hyperplane of equation

$$x_0 G'_0(x) + x_1 G'_1(x) + \cdots + x_n G'_n(x) = 0, \tag{23}$$

the *tangent hyperplane* to V at A.

Remark. By *Euler's formula* $d_0 G'_0(c) + \cdots + d_n G'_n(c) = dG(x)$, a point where all the partial derivatives of G vanish is on V if d is not a multiple of the characteristic of K.

If $G(x_0, \ldots, x_n)$ is obtained by homogenizing $F(y_1, \ldots, y_n)$, it is easy to see that $G'_i(1, y_1, \ldots, y_n) = F'_i(y_1, \ldots, y_n)$ for $i = 1, \ldots, n$, whence

$$G'_0(1, y_1, \ldots, y_n) = dF(y) - y_1 F'_1(y) - \cdots - y_n F'_n(y).$$

Applying this formula to a simple point $A = (1, a_1, \ldots, a_n)$ of V we see that, since $F(a) = 0$, equation (23) reduces to the affine equation for the tangent hyperplane (19).

Three important theorems.

We will often use phrases borrowed from elementary geometry, such as "draw the line passing through two points", "collinear points", "concurrent lines", "coplanar lines", and so on. The line passing through two (distinct) points of an affine or projective space will be denoted by D_{ab} or simply ab.

Theorem 5 (Desargues). *In a projective space P, let D, D' and D'' be distinct lines having a common point O. If $A, B \in D$, $A', B' \in D'$ and $A'', B'' \in D''$ are points distinct from one another and from O, the three intersection points $I = D_{AA'} \cap D_{BB'}$, $J = D_{AA''} \cap D_{BB''}$ and $K = D_{A'A''} \cap D_{B'B''}$ are collinear.*

Proof. These intersection points are well-defined: D and D', for example, lie on the same plane, so $D_{AA'}$ and $D_{BB'}$ also line on that plane; the two being distinct we can apply theorem 1 (section 1). To show collinearity, start with the case when the three lines D, D' and D'' are not coplanar. Then they generate a three-dimensional projective linear space, which contains the planes $AA'A''$ and $BB'B''$. Again by theorem 1, these two planes must have a line in common, which contains I, J and K.

The case when D, D' and D'' are coplanar follows by projection, but we will give a direct proof. In the plane of the three lines, let the line at infinity be D_{IJ}, and assume first that $O \notin D_{IJ}$. Let the origin be O. Looking at $A, B, \ldots, B''$ as vectors, we can find scalars $a, a', a'' \in K$ such that $B = aA$, $B' = a'A'$ and $B'' = a''A''$. Since I is at infinity, AA' and BB' are parallel, so there exists $c \in K$ such that $B' - B = c(A' - A)$, that

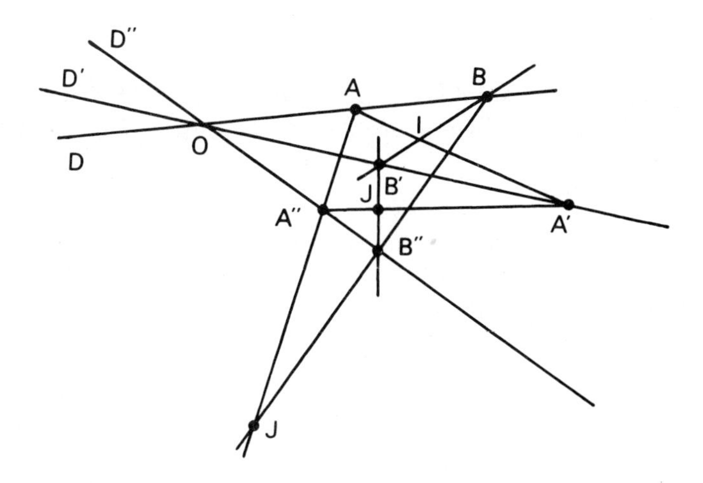

is, $a'A' - aA = cA' - cA$; this implies $a' = a = c$ because A and A' are linerarly independent. Similarly $a = a''$. But then $B'' - B' = a(A'' - A')$, which shows that $D_{A'A''}$ and $D_{B'B''}$ are parallel, that is, their intersection K is on the line at infinity D_{IJ}.

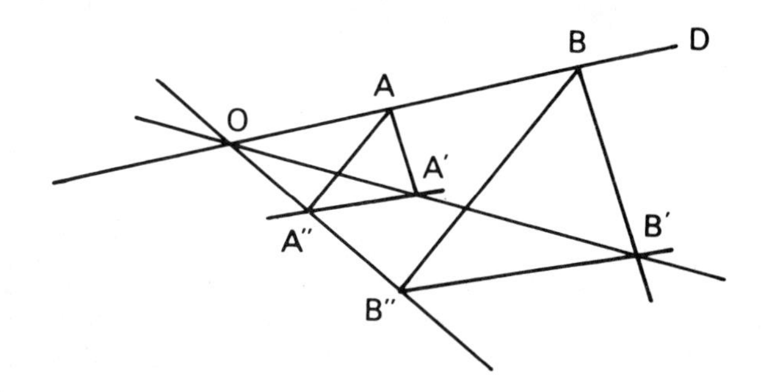

Finally, if $O \in D_{IJ}$, all three lines D, D' and D'' are parallel and $ABB'A'$ and $ABB''A''$ are parallelograms. The translation $B-A$ takes A' to B' and A'' to B'', so $D_{A'A''}$ and $D_{B'B''}$ are parallel, and again K is at infinity. □

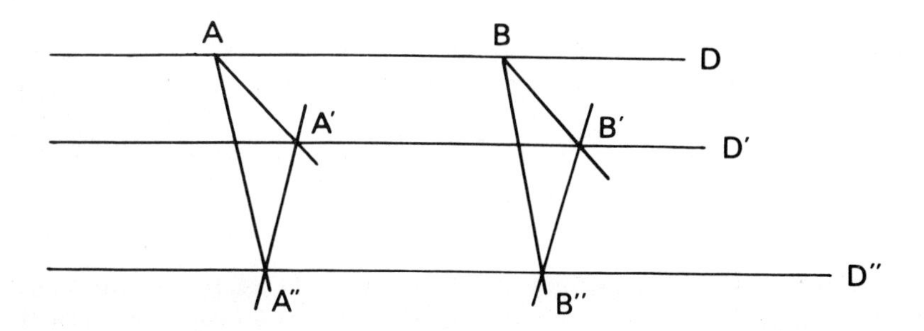

Theorem 6 (Pappus). *Let P be a projective plane over a field K. The following conditions are equivalent:*

(1) *For any two distinct lines D, D' and any points $A, B, C \in D$ and $A', B', C' \in D'$, all distinct, the points $I = D_{AB'} \cap D_{BA'}$, $J = D_{CA'} \cap D_{AC'}$ and $K = D_{BC'} \cap D_{CB'}$ are collinear.*

(2) *K is commutative.*

Proof. Notice first that in an affine plane two points with coordinates (p, q) and (p', q') are collinear with the origin I if and only if $p^{-1}q = p'^{-1}q'$.

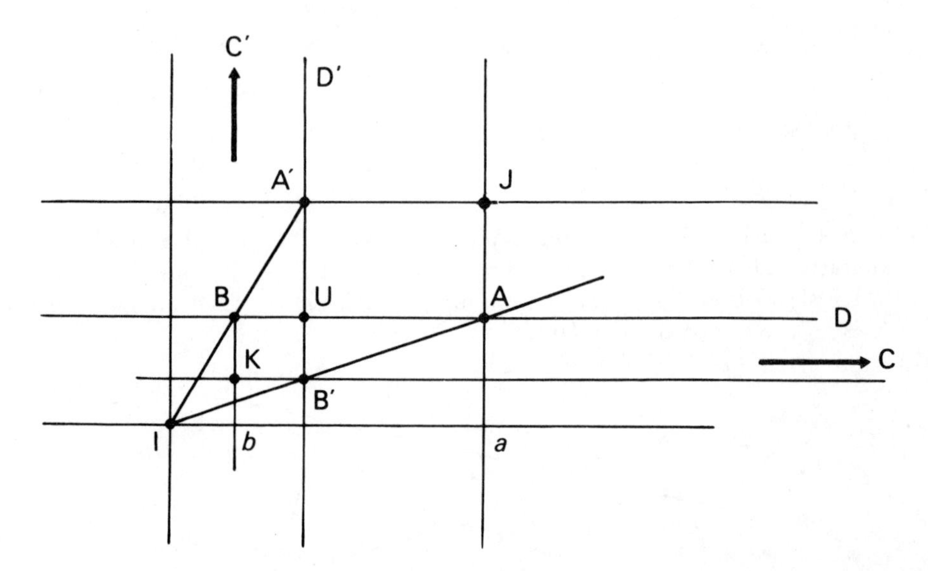

Now take I, C and C' as vertices of a projective frame, $D_{CC'}$ as the line at infinity and $U = D \cap D'$ as the unit point. All points in D have second coordinate 1; write $A = (a, 1)$ and $B = (b, 1)$. Since A, B' and I are collinear and B' has first coordinate 1, we have $B' = (1, a^{-1})$. Similarly, the coordinates of A' are $(1, b^{-1})$. Thus J has coordinates (a, b^{-1}) and K has coordinates (b, a^{-1}); they are collinear with the origin if and only if $a^{-1}b^{-1} = b^{-1}a^{-1}$, if and only if $ab = ba$. Since $a, b \neq 0$ are arbitrary I, J, K are always collinear if and only if K is commutative. □

Theorem 7 (fundamental theorem of projective geometry). *Let $P = \mathbf{P}(V)$ and $P' = \mathbf{P}(V')$ be two projective spaces of same dimension $n \geq 2$ over fields K and K'. If $f : P \to P'$ is a bijection that takes collinear points into collinear points, f is induced on P by a bijection $g : V \to V'$ that is additive and satisfies $g(ax) = s(a)g(x)$, where $s : K \to K'$ is a fixed field isomorphism.*

A bijection taking collinear points into collinear points is called a *collineation*. An additive map $g : V \to V'$ such that $g(x) = s(a)g(x)$ for some field isomorphism $s : K \to K'$ is called *semilinear* (with respect to to s). It is clear that a bijective semilinear map preserves linear dependence, and that it induces a map $f : \mathbf{P}(V) \to \mathbf{P}(V')$; thus the converse of theorem 7 is trivially true.

We could have restricted the theorem's hypothesis to sets of three collinear points, because for (fixed) $a, b \in A$ and a variable $x \in D_{ab}$, the image $f(x)$ is on the line $D_{f(a),f(b)}$.

Corollary. *A collineation from a projective space of dimension at least two into itself is a composition of an automorphism of the field of scalars with a projective transformation.* □

The fields $\mathbf{Q}$, $\mathbf{R}$ and $\mathbf{F}_p$, for p prime, have no non-trivial automorphisms, so for such field collineations and projective transformations are equivalent.

Proof of theorem 7. Let $a_0, \ldots, a_n$ be projectively independent points in P. For $j = 0, \ldots, n$, denote by L_j the projective linear space generated by $a_0, \ldots, a_j$ and by L'_j the projective linear space generated by $f(a_0), \ldots, f(a_j)$.

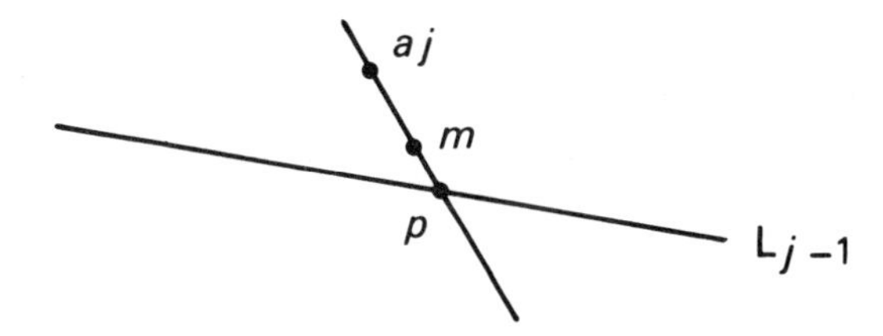

First we see, by induction on j, that $f(L_j) \subset L'_j$: indeed, for every $m \in L_j$, the line $a_j m$ intersects L_{j-1} at a point p; since $f(m)$ is collinear with $f(a_j)$ and $f(p)$, which lies in $L'_{j-1} \subset L'_j$, we obtain $f(m) \in L'_j$. This also implies that $f(a_0), \ldots, f(a_n)$ are projectively independent, because $P' = f(P) = f(L_n) \subset L'_n$ by surjectivity.

On the other hand we have $f(L_j) \supset L'_j$, because f is surjective, and points $m \in P \setminus L_j$ are mapped outside L'_j (complete the set $(a_0, \ldots, a_j, m)$ into a set of $n+1$ projectively independent points; by the previous paragraph $f(a_0), \ldots, f(a_j), f(m)$ will be projectively independent). Thus we conclude that $f(L_j) = L'_j$.

Now choose in P an origin O and a hyperplane at infinity H. Then $f(H)$ is a hyperplane H' of P', which we also take to be at infinity; and we take $O' = f(O)$ as the origin. In this way we get a bijection, also denoted by f, between the vector spaces $E = P \setminus H$ and $E' = P' \setminus H'$; we'll be done if we show that f is semilinear with respect to some field automorphism s, since then $g = f \times s$ will be the desired map from $V = E \times K$ into $V' = E' \times K'$.

The map $f : E \to E'$ takes lines into lines and parallel lines into parallel lines. Since $f(O) = O'$, the parallelogram rule shows that $f(x+y) = f(x) + f(y)$ when x and y are linearly independent.

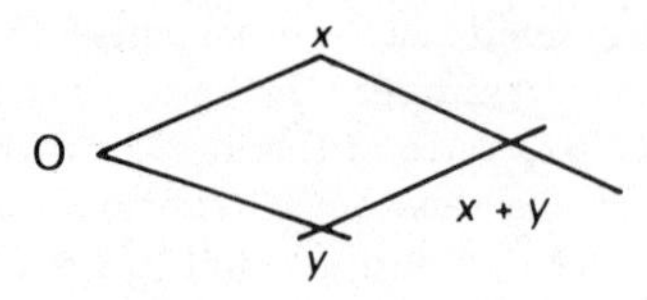

Otherwise we have $y \in Kx$ and, since we assumed $\dim E \geq 2$, we may we take a point $z \notin Kx$, which will be linearly independent of x, y and $x+y$. In addition, $y+z$ is linearly independent of x. In this case, too, the additivity of f is verified, because

$$f(x+y+z) = f\big((x+y)+z\big) = f(x+y)+f(z) = f\big(x+(y+z)\big)$$
$$= f(x)+f(y+z) = f(x)+f(y)+f(z).$$

For $a \in K$ and $x \neq O$, the points O, x and ax are collinear, hence so are O, $f(x)$ and $f(ax)$. Thus there exists $s(a,x) \in K'$ such that $f(ax) = s(a,x)f(x)$. If x and y are linearly independent, so are $f(x)$ and $f(y)$, and we see, by calculating $f\big(a(x+y)\big)$ in two ways, that $s(a,x) = s(a,y) = s(a,x+y)$; this can be checked for x and y linearly dependent as well, using an auxiliary vector z as in the previous paragraph. Thus $s(a,x)$ does not depend on x, and we denote it by $s(a)$.

For $x \neq O$, hence $h(x) \neq O'$, the formulas $f\big((a+b)x\big) = f(ax)+f(bx)$ and $f\big(a(bx)\big) = f\big((ab)x\big)$ immediately give $s(a+b) = s(a)+s(b)$ and $s(ab) = s(a)s(b)$. Since $f(Kx) = K'f(x)$, we conclude that $s: K \to K'$ is surjective, hence an isomorphism. □

Remark on the affine analogue of theorem 7.

Let $f: A \to A'$ be a bijection taking triples of collinear points into triples of collinear points, where A and A' are affine spaces of same dimension $n \geq 2$, over K and K', respectively. When $K = \mathbf{F}_2$, this assumption is vacuous, because lines have only two elements. However, if $K \neq \mathbf{F}_2$, one can show that f is semilinear (as a map from the vectorialization of A at an arbitrary point O to the vectorialization of A at $O' = f(O)$).

The proof is very similar to that of theorem 7. One takes affinely independent points $a_0, \ldots, a_n \in A$, and, denoting by L_j and L'_j the affine subspaces generated by $a_0, \ldots, a_j$ and $f(a_0), \ldots, f(a_j)$, respectively, one shows by induction over j that $f(L_j) \subset L'_j$. For $m \in L_j$, there is no difficulty if the line ma_j intersects L_{j-1}. If not, ma_j is parallel to some direction in L_{j-1}; one then takes an auxiliary point $p \in D_{ma_0}$, so the line a_jp is not parallel to L_{j-1}. Setting $q = a_jp \cap L_{j-1}$, one concludes from the collinearity of a_j, p and q that $f(p) \in L'_j$, and from the collinearity of m, p and a_0 that $f(m) \in L'_j$.

There follows from the surjectivity of f, as in theorem 7, that $f(L_j) = L'_j$ for every j, and that f takes lines into lines and parallel lines into parallel lines. An application of the parallelogram rule and the same calculations as in theorem 7 complete the proof.

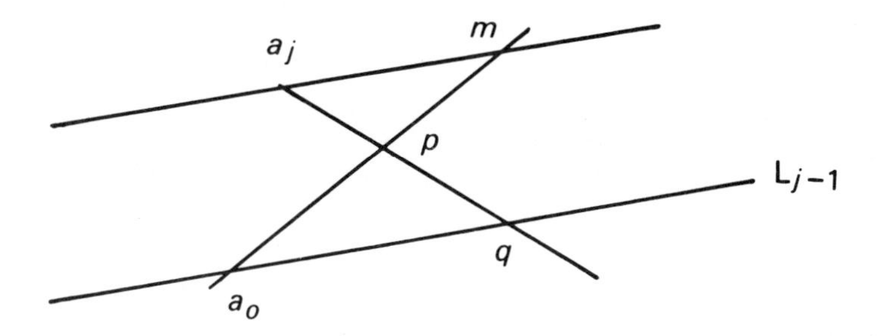

1.4. Axiomatic Presentation of Projective and Affine Planes

Incidence axioms: projective case

The fundamental theorem of projective geometry (theorem 7) hints that it is possible to reconstruct projective geometry from the notion of collinearity. That's just what we're going to do, axiomatically, in the case of the plane.

Consider a set P of points, called a *plane*, and a non-empty family of proper, non-empty subsets of P, called *lines*. Assume that the following *incidence axioms* are satisfied:

(A1) *Two distinct points in P belong to exactly one line.*
(A2) *Two distinct lines in P have exactly one common point.*

Remarks

(1) Notice the symmetry of the two assertions, which can be rephrased to say that "two points determine a unique line" and "two lines determine a unique point". We will come back to this topic in section 5, when we discuss duality. Notice also that A1 by itself already implies that two distinct lines have at most one common point, and similarly for A2.
(2) An axiomatic definition of n-dimensional projective spaces would involve $n-1$ families of non-empty, proper subsets of P, the j-dimensional projective linear subspaces of P for $j = 1, \ldots, n-1$, satisfying the following conditions: any $j+1$ points not contained in an $(j-1)$-dimensional projective linear space determine a unique j-dimensional projective linear space; any intersection of projective linear spaces is one; and the dimension of the intersection of two projective linear spaces is given by the formula in theorem 1, the notion of the projective linear space generated by a set making sense by the previous condition. This is all quite easy to write down explicitly in the case $n = 3$.

The following are immediate consequences of axioms A1 and A2. (We denote the line going through a and b by D_{ab} or ab.)

(1) *Since P has non-empty proper subsets, P has at least two points.* Furthermore, since all lines are distinct from P, A1 implies that P has at least three points.

(2) *No line consists of only one point.* Otherwise, by A2, every line would go through that point, say a; taking distinct points $b, c \neq a$ we'd have $ab = bc$ (by A1), so ab would actually be the whole plane.

(3) *Is the union $D \cup D'$ of two lines distinct from P?* Denote by a the common point of D and D'. If there are two points $b, c \in D$ distinct from a, and two points $b', c' \in D'$ distinct from a, the common point of bb' and cc' is outside D and D'. Thus $P = D \cup D'$ can only happen when D, for example, has exactly two points a and b, and D' is the complement of b. Such a plane is declared uninteresting and excluded from further consideration.

. b

a . . . . . D′

(4) *Any two lines D and D' are in bijection with one another.* Take $p \notin D \cup D'$ and project D onto D' through p, that is, associate with each $m \in D$ the unique point $m' \in D'$ where pm intersects D'. This map is clearly invertible, hence a bijection.

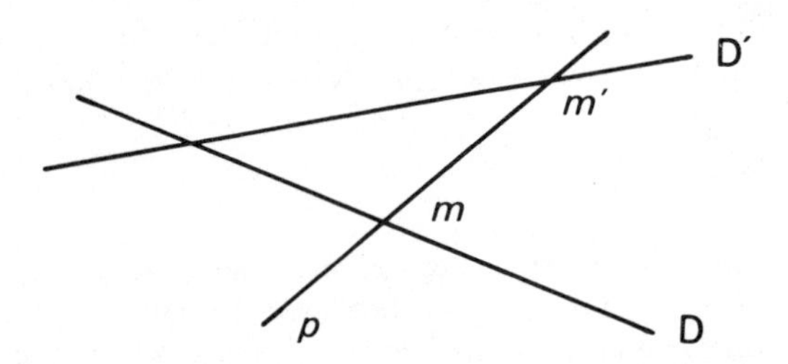

(5) *If the lines have finite cardinality $q + 1$, the plane contains $q^2 + q + 1$ points and the same number of lines.* Take a line D and a point $a \notin D$. All lines in a are in bijection with the points in D, so there are $q + 1$ of them. Each line contains q points distinct from a, whence $\#(P) = (q + 1)q + 1$. In order to count the lines we notice that, except for D, there are q lines going through each of the $q + 1$ points of D, hence a total of $1 + q(q + 1)$ lines.

(6) *We have $q \geq 2$, and consequently $\#(P) \geq 7$.* If $q = 1$ all lines have two points and all two-point sets in P are lines (by A1); this can't happen if $\#(P) \geq 4$ because we'd have disjoint lines, contradicting A2, and if $\#(P) = 3$ it means the plane is uninteresting.

A projective plane is said to be *desarguesian* if it satisfies the following axiom:

(A3) (*Desargues's axiom*). *For any distinct concurrent lines D, D', D'' and any points $a, b \in D$, $a', b' \in D'$ and $a'', b'' \in D''$, distinct from one another and from the common point of D, D', D'', the points $D_{aa'} \cap D_{bb'}$, $D_{aa''} \cap D_{bb''}$ and $D_{a'a''} \cap D_{b'b''}$ are collinear.*

Incidence axioms: affine case

Let P be a (not uninteresting) projective plane. Let $A = P \setminus D_0$ be the complement of a line D_0 of P, which will be called the line at infinity. Call a subset of A a *line* if it is the intersection of A with a line of P distinct from D_0; the lines of A are non-empty, by consequence (2) above, and distinct from A, by consequence (3). Call two lines of A *parallel* if the lines of P that they come from meet at infinity. Axiom A1 becomes:

(A′1) *Two distinct points in A determine a unique line.*

It follows that two distinct lines in A have at most one common point. By A2, two lines without a common point are parallel. Applying A1 to a point in A and a point at infinity (that is, in D_0), we get:

(A′2) *Through each point in A we can draw a unique line parallel to a given line.*

If $q+1$ is the cardinality of the lines of P, all the lines in A have q elements (recall that $q \geq 2$, by consequence (6) above). The affine plane A contains q^2 points and q^2+q lines.

Conversely, we call an *affine plane* any set A of points with a non-empty set of non-empty, proper subsets, called lines, that satisfy axioms A′1 and A′2. It follows from A′1 that two distinct lines have at most one common point. We say that two lines are *parallel* if they coincide or have no points in common. Parallelism is an equivalence relation: reflexivity and symmetry are obvious, and, if D is parallel to D' and to D'', either D' and D'' are disjoint, or, if they have a common point a, they coincide (by A′2 applied to a and D). A *direction* is an equivalence class of lines; let D_0 be the set of directions.

By adjoining D_0 to A we obtain a set P. By definition, the *lines* of P are D_0 and the subsets $\hat{D}$ obtained by adding to each line of A its direction (also called its point at infinity); these subsets are all non-empty and distinct form P. One sees immediately that P is a projective plane: A1 follows from A′1 and A′2, and A2 from the fact that two disjoint lines in A share a point at infinity. Furthermore this projective plane is not uninteresting.

Otherwise it contains a line E whose complement is a point a, by (3) above. We have $E \neq D_0$ because A, which contains non-empty, proper subsets, cannot have only one point. Thus E comes from a line D of A; let b be a point in D. The point at infinity of ab lies outside E; furthermore it cannot be a, otherwise $A = D$. This gives two points outside E, a contradiction.

Thus, if we disregard uninteresting planes, the study of affine planes and that of projective planes are interchangeable. This situation is analogous to the one encountered in section 3.

An affine plane is called *desarguesian* if it satisfies the following axiom (a particular case of A3):

(A′3) *Let D, D', D'' be distinct concurrent or parallel lines in A, and $a, b \in D$, $a', b' \in D'$, $a'', b'' \in D''$ points distinct from one another and from the common point of D, D', D'', if it exists. If $D_{aa'}$ is parallel to $D_{bb'}$ and $D_{aa''}$ is parallel to $D_{bb''}$, then $D_{a'a''}$ is parallel to $D_{b'b''}$.*

The fundamental theorem

We say that two projective (or affine) planes are *isomorphic* if there exists a bijection between the two that maps the family of lines of the first onto the family of lines of the second.

Theorem 8.

(a) *Every desarguesian projective plane P is isomorphic to a plane $P(V)$, where V is a three-dimensional vector space over a field.*

(b) *Every desarguesian affine plane E is isomorphic to an affine plane over a field.*

Proof. It is enough to show (b), because, if P is a projective plane and D_0 is a line in P, we can apply (b) to $P \backslash D_0$, which gives (a) by the considerations in section 3. Notice that, because of theorem 5, the particular case A′3 of Desargues's axiom implies the more general axiom A3.

The proof of part 8(b) is long. It will comprise the following steps:

- a direct analysis of the "small" affine planes ($q = 2$ or 3), where the difficulties arising from the collinearity of certain points in the proof are not as easy to get around as when the plane is bigger;
- the definition of an equivalence relation on pairs of points of E, the equivalence classes of which are called vectors;
- the definition of a group structure on the set V of these equivalence classes and of a group action of V on E by so-called translations;
- the definition of certain endomorphisms of V, called homotheties, and the proof that they form a field K.

The small planes

If $q = 2$, E has four points, and lines are just two-element subsets. Thus E is isomorphic to the affine plane over $\mathbf{F}_2$.

If $q = 3$, choose two intersecting lines D and D', and name the points of each 0, 1 and 2, where 0 is common to both. Consideration of the six lines parallel to either D or D' shows that we can list the points of E by the pairs (i, j), for $i, j = 0, 1, 2$.

2 . . .
1 . . .
0 . . .
0 1 2

Any other line intersects each line parallel to either D or D' in a unique point, so its points are of the form $(i, s(i))$, where s is an element of the symmetric group S_3, that is, a permutation of $\{0,1,2\}$. Since there are six such lines—because $q^2 + q = 12$—the set of lines parallel to neither D nor D' is in one-to-one correspondence with S_3. But this is exactly the situation for $\mathbf{F}_3 \times \mathbf{F}_3$: the lines non-parallel to the coordinate axes are of the form $y = ax + b$, for $a, b \in \mathbf{F}_3$ and $a \neq 0$, and so correspond also to the six permutations of $\mathbf{F}_3 = \{0,1,2\}$.

The equivalence relation

Given two distinct, non-collinear points a, b, a', b', we set

(I) $$(a,b) \sim (a',b')$$

if $D_{a'b'}$ is parallel to D_{ab} and $D_{bb'}$ is parallel to $D_{aa'}$. The relation $(a,b) \sim (a',b')$ is equivalent to $(b,a) \sim (b',a')$, to $(a,a') \sim (b,b')$ and to $(a',a) \sim (b',b)$. It is also equivalent to $(a',b') \sim (a,b)$, that is, $\sim$ is a symmetric relation. Any three points determine the fourth.

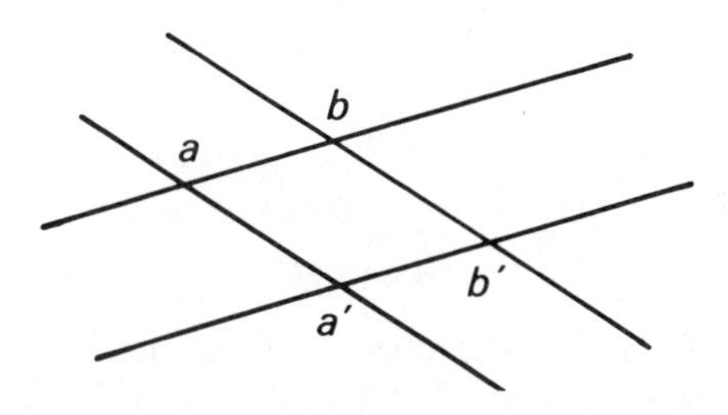

Lemma 1. *If the lines D_{ab}, $D_{a'b'}$ and $D_{a''b''}$ are distinct, $\sim$ is transitive: $(a,b) \sim (a',b')$ and $(a',b') \sim (a'',b'')$ imply $(a,b) \sim (a'',b'')$.*

Proof. This follows from Desargues's axiom A′3: $D_{aa'}$ parallel with $D_{bb'}$ and $D_{a'a''}$ parallel with $D_{b'b''}$ imply that $D_{aa''}$ is parallel with $D_{bb''}$. □

Now, if a, b, a', b' are on the same line D, with $a \neq b$, we set

(II) $$(a,b) \sim (a',b')$$

if there exist points $u, v \notin D$ such that $(a,b) \sim (u,v)$ and $(u,v) \sim (a',b')$.

This relation is reflexive and symmetric. If u is given, v is uniquely determined.

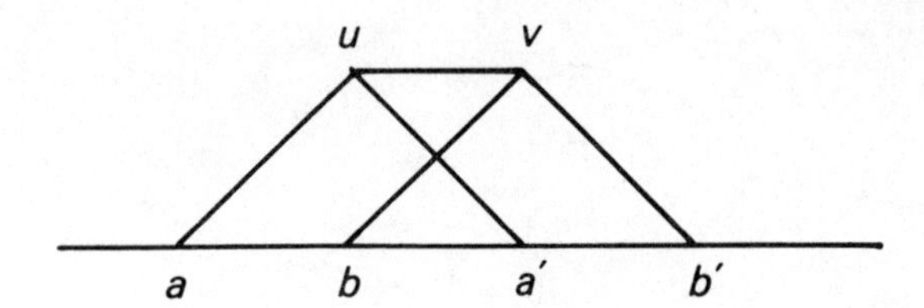

Lemma 2. *If a, b, a', b' are collinear, the relation $\sim$ defined by (II) is independent of u.*

Proof. Assume we also have $(a,b) \sim (u',v')$ and $(u',v') \sim (a',b'_1)$. If u' is outside D and D_{uv}, we apply lemma 1 and obtain $b'_1 = b'$. Otherwise we take u'' outside D and D_{uv}, which is possible because $q \geq 3$, and define v'' by $(u'',v'') \sim (u,v)$; an application of lemma 1 to (a,b), (u,v), (u'',v'') and to (u'',v''), (u',v'), (a',b'_1) shows again that $b' = b'_1$. □

Corollary. *For any given points a, b, a' on a line D, there exists a unique $b' \in D$ such that $(a,b) \sim (a',b')$.* □

Lemma 3. *The relation $\sim$ between pairs (a,b) and (a',b') of distinct points($a \neq b$, $a' \neq b'$) defined by (I) and (II) is an equivalence relation. Any three of the points determine the fourth.*

Proof. Symmetry has been shown. Reflexivity follows form lemma 1 if D_{ab}, $D_{a'b'}$ and $D_{a''b''}$ are distinct; otherwise we call in an extra pair (u,v) lying on a parallel line distinct from the previous three (again possible because $q \geq 3$). □

Finally, for $a = b$, we set

$$(a,a) \sim (a',b')$$

if and only if $a' = b'$. This completes the definition of the relation $\sim$ on the set of ordered pairs (a,b); it is clear that $\sim$ is and equivalence relation.

Vectors and translations

A *vector* is a $\sim$-equivalence class of ordered pairs of points in E. The class of a pair (a,b) is denoted by $b-a$, and we say that (a,b) is a *representative* of $b-a$. A vector x has a unique representative (a,b) with given origin a, since any point is determined by the other three in the relation $(a,b) \sim (a',b')$. The class of pairs (a,a) is called the zero vector, and is denoted by 0. For all representatives (a,b) of a given non-zero vector, the lines D_{ab} are parallel; their direction is called the direction of x.

The *sum* of two vectors x, y is the vector z defined as follows: take a point a, let (a,b) be the representative of x with origin a and (b,c) the representative of y with origin b, and set $z = c - a$. To show that z is independent of a, consider this construction for some other point a': the lines $D_{aa'}$, $D_{bb'}$ and $D_{cc'}$ are parallel, and, if $D_{aa'}$ is not parallel to D_{ab}, D_{ac}, or D_{bc}, Desargues's axiom A'3 shows that $c' - a' = c - a$. Otherwise we introduce an auxiliary point a'': since $q \geq 4$, the plane has five distinct directions, and we can choose two that are not those of D_{ab}, D_{ac}, or D_{bc}, and make them the directions of aa'' and $a''a'$.

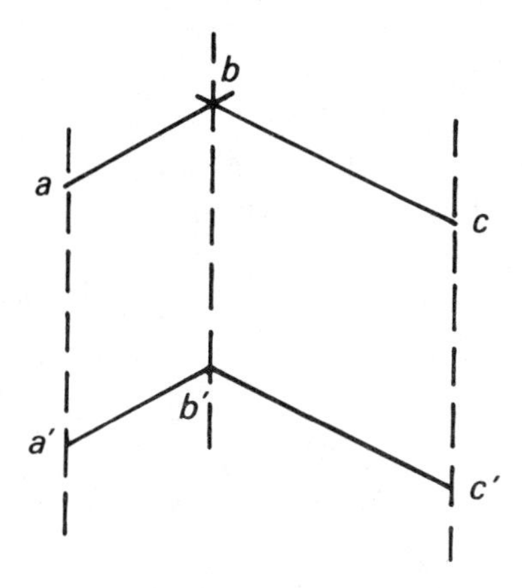

The sum of two vectors x and y will be called $x + y$. Clearly $+$ is associative. To show commutativity, we use the parallelogram rule (I) and the definition of $\sim$ when the directions of x and y are distinct; otherwise we set $y = y' + y''$, where the directions of y' and y'' are distinct from that of x, and write

$$\begin{aligned} x + y = x + (y' + y'') &= (x + y') + y'' = (y' + x) + y'' \\ &= y' + (x + y'') = y' + (y'' + x) = y + x. \end{aligned}$$

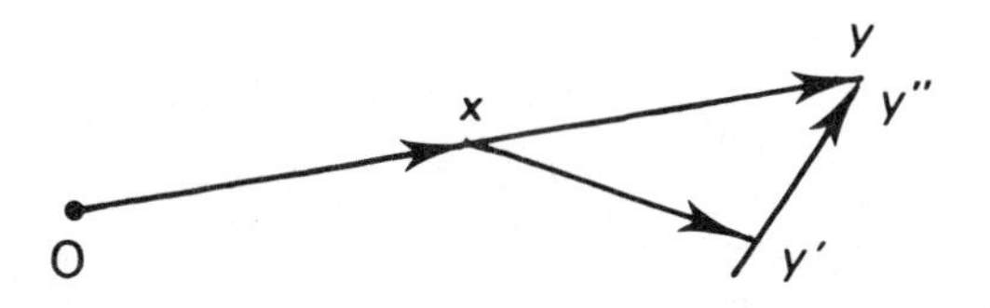

The zero vector is obviously the identity element, and $b - a$ is the negative of $a - b$. This shows that the set V of vectors is a commutative group.

Lemma 4. *The relations $b - a = b' - a'$ and $a' - a = b' - b$ are equivalent.*

Proof. This has been seen when the four points a, b, a', b' are not collinear. In the general case, $b - a = b' - a'$ implies, by the definition of the sum of two vectors and its commutativity:

$$a' - a = (a' - b) + (b - a) = (b' - a') + (a' - b) = b' - b. \qquad \square$$

Finally, let x be a vector. For every point $m \in E$, denote by $T_x(m)$ the unique point m' such that $m' - m = x$. By the definition of the sum of two vectors, we have $T_y(T_x(m)) = T_{x+y}(m) = T_{y+x}(m)$; in addition we have $T_x = \mathrm{Id}_E$ if and only if $x = 0$. Thus each T_x is a bijection of E, which we call the *translation* by x. The discussion above shows that V acts on E by translations, and that this action is simply transitive: the unique translation that takes a to b is T_{b-a}.

The field of homotheties

Choose a point $O \in P$, and identify each point $a \in E$ with the vector $a - O$. This identifies E with the commutative group V, the sum of two points non-collinear with O being given by the parallelogram rule. Notice that every line going through O is a subgroup of E, because the sum of two vectors of same direction has the same direction as the two.

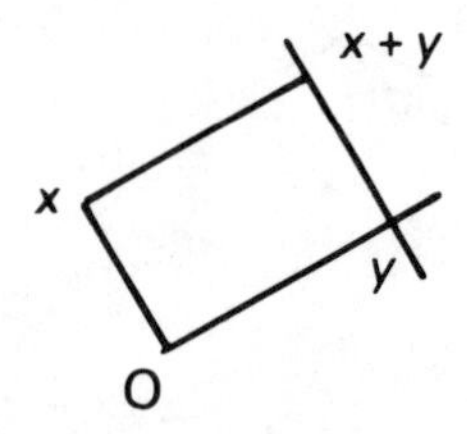

Now fix a line A through O and a point 1 on A, distinct from O. For every $a \in A$ and every $x \notin A$, consider the parallel to D_{1x} going through a; it intersects the line Ox at a point which we call $h_a(x)$. We have $h_0(x) = O$ and $h_1(x) = x$ for every $x \notin A$. Temporarily fix $a \neq O$ on A and set $h = h_a$.

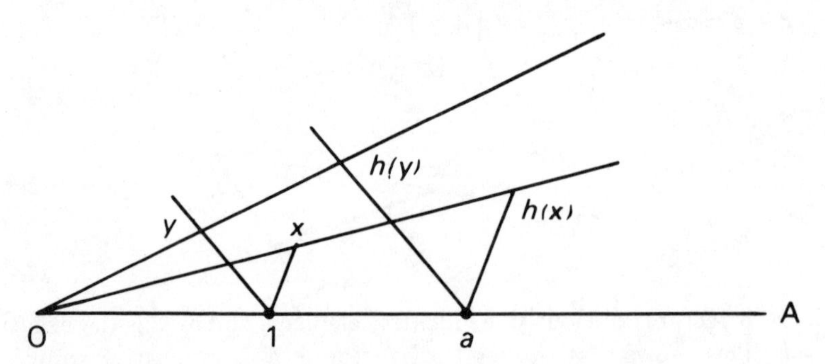

Lemma 5. *If $x, y \notin A$ are distinct, $D_{h(x)h(y)}$ is parallel to D_{xy}.*

Proof. This is obvious if D_{xy} contains O or 1. Otherwise it follows from Desargues's axiom A′3. □

Corollary. *If $D \subset E$ is any line, h transforms points on D into points on a line parallel to D, with the possible exception of points in $D \cap A$.* □

Lemma 6. *If x, y and $x+y$ lie outside A, we have*

$$h(x+y) = h(x) + h(y).$$

Proof. If x and y are not collinear with O, we apply lemma 5 and the parallelogram rule. If y is on $D = D_{0x}$, we take a point $u \notin D$ such that $u, x+y+u, y+u \notin A$ (this is possible because $q \geq 4$: we must avoid D and three parallels to A, so we take u on a fourth parallel to A, and avoid its intersection with D). Then, since $x, y, x+y \in D$, we have $y+u \notin D$, $x+y+u \notin D$, which gives

$$h(x+y+u) = h(x+y) + h(u),$$
$$h(x+y+u) = h(x) + h(y+u) = h(x) + h(y) + h(u),$$

as we wished to prove. □

We will now define $h(x)$ for $x \in A$. Take $u \notin A$ and consider $d(u) = h(x+u) - h(u)$. For $v \notin A$, the difference $d(u) - d(v)$ equals $h(x+u) - h(v) = h(x+v) - h(u)$, which is zero if $x+u+v \notin A$, since both sides equal $h(x+u+v)$. If $x+u+v \in A$ we use an auxiliary point w such that $x+u+w \notin A$ and $x+v+w \notin A$; we just have to avoid two parallels to A. This shows that $d(u)$ is independent of u, and we set

(III) $\qquad h(x) = h(x+u) - h(u) \qquad$ for any $u \notin A$.

Lemma 7. *The map h thus defined is an endomorphism of the group E.*

Proof. The relation $h(x+y) = h(x) + h(y)$, which holds outside A by lemma 6, still holds by (III) if one of the points is on A. If x, y and $x+y$ all lie in A we use the old trick: computing $h(x+y+u)$ in two ways, for $u \notin A$. □

Lemma 8. *We have $h_a(x) + h_b(x) = h_{a+b}(x)$ for every $a, b \in A$ and every $x \in E$.*

Proof. By bringing in an auxiliary point if necessary, we can assume $x \notin A$. By construction, $h_a(x) - a$ and $h_b(x) - b$ have the same direction as $x - 1$, hence so does their sum $h_a(x) + h_b(x) - (a+b)$. Thus $h_a(x) + h_b(x)$ is the intersection of D_{0x} with the parallel to D_{1x} going through $a+b$; but this intersection is $h_{a+b}(x)$. □

Lemma 9. *Let $a \in A$. Then h_a is the unique endomorphism of E taking every line into a parallel line and taking 1 to a. If a is non-zero, h_a is an automorphism.*

Proof. We first show that if such an endomorphism exists it is unique. If $x \notin A$, its image $h_a(x) = y$ is uniquely determined as the intersection of D_{0x} with the parallel to D_{1x} through a, because D_{0x} is invariant under h_x. For $x \in A$, uniqueness follows from (III).

Now we show that h_a has the right properties. We have seen that, if $D \neq A$ is a line through O, the image of D is contained in D; hence for every parallel $u+D$ to D, we have $h_a(u+D) = h_a(u)+h_a(D) \subset h_a(u)+D$. The corollary to lemma 5 shows that, for every parallel $v + A$ to A, with $v \notin A$, the image $h_a(v + A)$ is contained in a parallel to $v + A$, which is then necessarily $h_a(v) + A$. Since h_a is an endomorphism, we deduce that $h_a(A) \subset A$, showing that the image of any line is a line parallel to it.

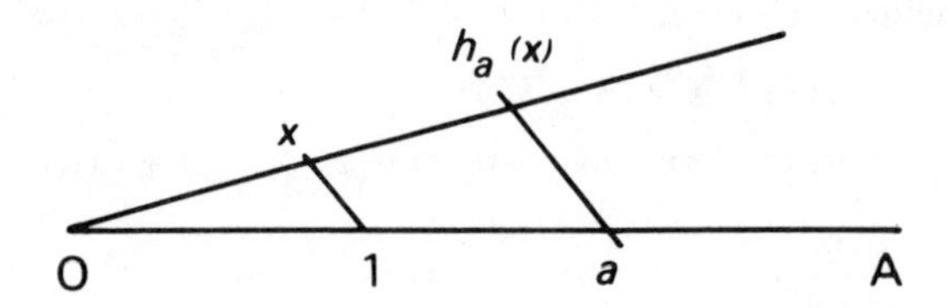

In particular, for $x \notin A$, the image $h_a(D_{1x})$ is contained in the parallel to D_{1x} going through $h_a(x)$. This parallel contains $h_a(1)$ and, by the definition of $h_a(x)$, also a; this, together with $a \in A$, implies $h_a(1) = a$.

Finally, if $a \neq 0$, every point $y \in E \setminus A$ is in $h_a(E)$, being the image of the point x where the parallel to D_{ay} through 1 intersects D_{0y}. By using an auxiliary point we see this is true for $y \in E$, that is, h_a is surjective, and maps every line *onto* a parallel line. To show injectivity, notice first that a point $x \notin A$ maps outside the origin, by construction; and a point $0 \neq u \in A$, in turn, maps to the intersection of A with the parallel to D_{xu} passing through $h_a(x)$, which is non-zero. □

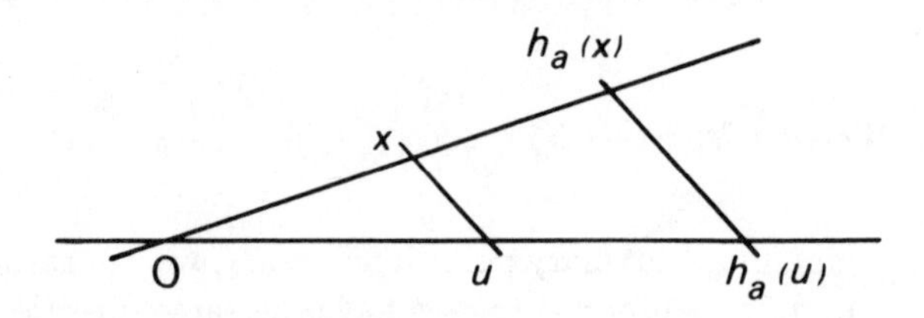

We're now practically done. For a and b in A, the composition $h_a \circ h_b$ is an endomorphism of E which maps each line into a parallel line; by lemma 9, $h_a \circ h_b = h_c$ for some $c \in A$. Set $ab = c$; this clearly defines a group law on A, with 1 as the identity. Since $ab = h_{ab}(1) = h_a\big(h_b(1)\big) = h_a(b)$, we get $a(b + b') = ab + ab'$ by the additivity of h_a. From lemma 8 if follows that $(a+a')b = ab+a'b$, so that A is a ring. Finally, for $a \neq 0$, the automorphism h_a^{-1} takes every line into a parallel line, whence, by lemma 9, $h_a^{-1} = h_{a'}$ for some $a' \in A$; since $aa' = a'a = 1$, we have shown that A is a field.

The formulas above show that, setting $ax = h_a(x)$ for $a \in A$ and $x \in E$, we make E a left vector space over A. Every line D through O is a one-dimensional vector subspace, because, if we fix a non-zero $u \in D$, the map $a \mapsto au$ is a bijection between A and D (the projection parallel to D_{u1}). By the parallelogram rule, two points of E non-collinear with O form a basis for E, which is consequently two-dimensional over A. As translates of lines through O, the lines of E are simply the one-dimensional affine spaces of E. This completes the proof of theorem 8. □

Comments on Desargues's axiom

(1) Desargues's axiom is independent of axioms A′1 and A′2. Here's an example of an non-desarguesian affine plane: Let a "line" in $\mathbf{R}^2$ (notice the quotation marks) be either

- an ordinary line of slope zero, infinity or negative;
- a broken line of the form

$$y = \begin{cases} m(x-t) & \text{for } x \leq t, \\ 2m(x-t) & \text{for } x \leq t \end{cases} \qquad (m > 0). \tag{IV}$$

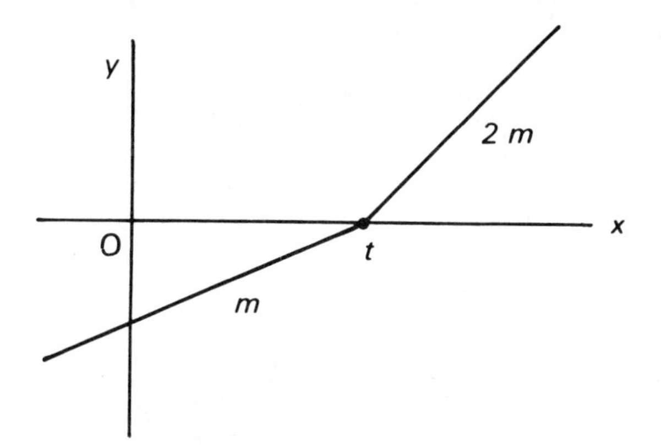

For $a, b \in \mathbf{R}^2$, axiom A′1 is clearly satisfied if the slope of the segment ab is zero, infintity or negative. If its slope is positive, let (x', y') and (x'', y'') be the coordinates of a and b, respectively. If y and y' have the same sign, A′1 is still satisfied: draw the line joining a and b as far as the x-axis, then extend it to the other half-plane by doubling or halving its slope. Finally, for $y' < 0$ and $y'' > 0$, (IV) can only be satisfied if $y''/(x''-t) = 2y'/(x'-t)$ (since $x' < x''$), and this equation has a unique solution

$$t = \frac{x'y'' - 2y'x''}{y'' - 2y'}.$$

To verify Euclid's axiom A′2, we can limit ourselves to the case of broken "lines". If D, D' are two such "lines", given by (m, t) and

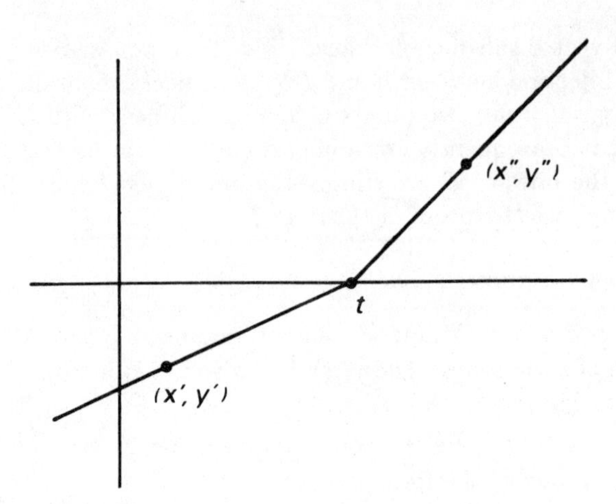

(m', t') in (IV), any intersection point (x, y) of D and D' must satisfy $m(x - t) = m'(x' - t')$ if $y \leq 0$, and $2m(x - t) = 2m'(x' - t')$ if $y \geq 0$. Both of these equations reduce to

$$(m - m')x = mt - m't'.$$

Thus D and D' have a unique common point if $m \neq m'$, and they are parallel if $m = m'$, that is, if they are obtained from one another by a translation along the x-axis. This clearly implies A′2.

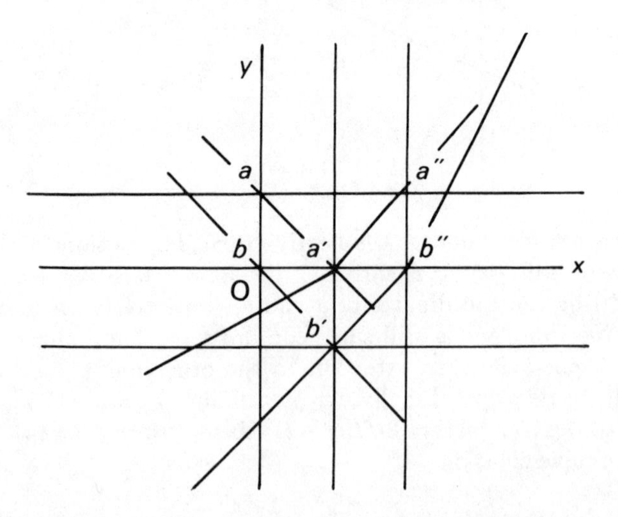

Finally, we show that this plane is not desarguesian. Denote the unique "line" through a and b by D'_{ab}. Consider the points $a = (0, 1)$, $b = (0, 0)$, $a' = (1, 0)$, $b' = (1, -1)$, $a'' = (2, 1)$ and $b'' = (2, 0)$, lying on the vertical "lines" $D = D_{ab}$, $D' = D_{a'b'}$ and $D'' = D_{a''b''}$. One immediately sees that $D'_{aa'}$ and $D'_{bb'}$ are ordinary and parallel, and so are $D'_{aa''}$ and $D'_{bb''}$; but $D'_{a'a''}$ and $D'_{b'b''}$ cannot be parallel, since they have different values of m in (IV).

(2) The study of finite (affine or projective) planes is connected with questions of statistics and combinatorics. Let q be the cardinality of the affine lines of a plane. If the plane is desarguesian, q is the order of a finite field (theorem 8), hence $q = p^r$ for some prime p. Since every finite field is commutative, Pappus's theorem (theorem 6) is satisfied.

There exist finite non-desarguesian fields. For instance, for p an odd prime and $r \geq 2$, consider on $\mathbf{F}_{p^r}$ the *twisted multiplication*

$$x \circ y = \tfrac{1}{2}\left(u(x)u(y)^p + u(x)^p u(y)\right),$$

where $u(x) = \frac{1}{2}(x + x^p)$. This multiplication law, which is bi-additive, makes $\mathbf{F}_{p^r}$ into a skew algebra, which yields a non-desarguesian projective plane. There is an analogous construction for $q = 2^r$ with $r \geq 4$. For q prime and $q = 4, 8$ all projective planes are desarguesian.

Little is known about other values of q. Combinatorial arguments show that there is no projective plane with $q = 6, 14, 21, 22$. The case $q = 10$ seems to be still open.

For more details see [Ha, chapter 20] and [Pi].

(3) If a projective plane can be embedded, as a projective linear space, into a higher-dimensional projective space, it must be desarguesian. Indeed, every plane Desargues configuration in the plane can be seen (cf. the proof of theorem 5) as the "projection" of a configuration in space, where the points I, J and K necessarily lie on the line common to the planes $AA'A''$ and $BB'B''$. The converse is clear by theorem 8.

(4) Another characterization of desarguesian planes is by saying that they have enough automorphisms (or *collineations*). For example, consider the following assertion:

(A″3) *For any line D and any three collinear points $O, u, v \notin D$ such that $u \neq O$ and $v \neq O$, there exists an automorphism of P fixing O and D(pointwise) and taking u to v.*

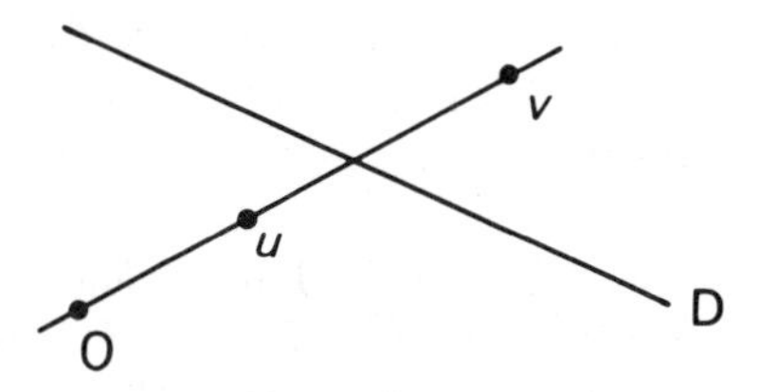

This is true for desarguesian planes: if D is sent to infinity and O is the origin, $P \setminus D$ becomes a vector space over a field K (theorem 8), and the desired automorphism is just the homothety $f(x) = ax$, for

$a \in K$, that takes u into v. (This homothety is semilinear only if K is non-commutative, because, for $b \in K$, we have $f(bx) = (aba^{-1})f(x) = s(b)f(x)$, where s is an inner automorphism of K.)

Conversely, if P satisfies A″3 and E, E', E'' are concurrent lines in P with points $u, v \in E$, $u', v' \in E'$ and $u'', v'' \in E''$ as in the hypothesis of axiom A3, let O be the common point of E, E', E'' and D the line joining $I = D_{uu'} \cap D_{vv'}$ with $J = D_{uu''} \cap D_{vv''}$. The automorphism f that takes u into v takes $D_{uu'}$ into $D_{vv'}$, since it leaves I fixed; since, for every $x \in P$, the points $O, x, f(x)$ are collinear, we see that $f(u')$ is equal to $v' = E' \cap D_{vv'}$. Similarly, $f(u'') = v''$, so f takes $D_{u'u''}$ to $D_{v'v''}$. This implies that the third intersection point $K = D_{u'u''} \cap D_{v'v''}$ is fixed, and consequently lies on D_{IJ}, because $K \neq 0$.

An elegant axiomatization of affine planes, based on the incidence axioms A′1 and A′2 and on the existence of enough automorphisms, was developed by Emil Artin [Ar].

1.5. Projective Spaces of Hyperplanes and Duality

Linear systems of hyperplanes

Recall that, if E is a (left) vector space over K, the set E^* of linear forms on E is a (right) vector space over K, called the dual of E. In finite dimension we have

$$\dim E^* = \dim E. \tag{24}$$

To every linear form f associate its kernel $\ker f$. If f is non-zero, $\ker f$ is a hyperplane of E. Two linear forms f, f' have the same kernel if and only if they are proportional, that is, $f' = fc$ for some $c \in K$. Thus the set of hyperplanes of E (or of $\mathbf{P}(E)$, which is the same), can be identified with the projective space $\mathbf{P}(E^*)$. We will assume $\mathbf{P}(E)$ has finite dimension n.

If we're given a projective coordinate system on $\mathbf{P}(E)$—coming, say, from a basis $\mathcal{B}$ of E—the hyperplane of equation $x_0u_0 + \cdots + x_nu_n = 0$, considered as a point in the projective space $\mathbf{P}(E^*)$ with the homogeneous coordinates coming from the basis of E^* dual to $\mathcal{B}$, has coordinates $(u_0, \ldots, u_n)$.

A projective linear space S of $\mathbf{P}(E^*)$ is called a *linear system of hyperplanes*, or a *pencil of hyperplanes* if $\dim S = 1$.

Theorem 9. *Given a linear system S of hyperplanes of $\mathbf{P}(E)$, there exists a projective linear space $B(S)$, called the base of S, such that S is the set of hyperplanes containing $B(S)$. We have* $\dim B(S) = \dim \mathbf{P}(E) - \dim S - 1$.

Proof. We think in terms of vector spaces. Recall that, if W is a vector subspace of a vector space V, the set W' of linear forms that vanish on W is a subspace of V^*, which can be identified with the dual of V/W;

hence $\dim W' = \operatorname{codim} W$ by (24). We apply that to the subspace T of E^* from which S comes. Observe that each element $x_0 \in E$ defines a linear form on E_0 (the evaluation map), so that, by (24), E can be identified with the dual of E^*; this is sometimes called *biduality*. Thus the subspace T' of E is the set of zeros common to the linear forms $f \in T$, that is, the intersection of their kernels. The set $(T')'$ of linear forms that vanish on T' obviously contains T; since $\dim(T')' = \operatorname{codim} T' = \dim T$, we have $(T')' = T$. Setting $B(S) = p(T')$, we see that the hyperplanes in S are exactly those containing $B(S)$. Since

$$\dim B(S) = \dim T' - 1 = \operatorname{codim} T - 1 = \dim E - \dim T - 1,$$

the formula about dimensions is proven. □

Examples. A pencil of hyperplanes is formed by all hyperplanes containing a projective linear space of codimension 2. In a three-dimensional space, a pencil of planes is the set of planes that pass through a given line; a two-dimension linear system of planes is the set of planes that pass through a given point. In the plane, a pencil of lines is formed by all lines going through a point; in the affine plane, we can talk about pencils of concurrent lines and pencils of parallel lines, those whose base point is at infinity.

Duality

Every theorem on projective linear spaces applies, of course, to linear systems of hyperplanes. To the notion of the projective linear space generated by a set of points a_i there corresponds the notion of the smallest linear system containing the hyperplanes H_i, which is, by theorem 9, the intersection of the H_i. Conversely, to an intersection of projective linear spaces there corresponds an intersection of linear systems S_i and the base of this intersection is the projective linear space generated by the bases $B(S_i)$, again by theorem 9. Thus we can establish a "dictionary" to translate to the language of linear systems of hyperplanes all general statements about projective linear spaces. Here is the dictionary in the case of the plane:

Point	Line
Line	Pencil of lines, or its base point
Collinear points	Concurrent lines
Line containing two points	Intersection of two lines

Because of biduality, this table can be read both ways.

As an example, let's translate the theorems of Pappus and Desargues (theorems 5 and 6):

Let D, D', D'' be concurrent lines.	Let a, a', a'' be collinear points.

Take points $a, b \in D$, $a', b' \in D'$ and $a'', b'' \in D''$.

Let i (resp. j, k) be the intersection point of the lines aa' and bb' (resp. aa'' and bb'', $a'a''$ and $b'b''$).

Then i, j, k are collinear.

Take lines $D, E \ni a$, $D', E' \ni a'$ and $D'', E'' \ni a''$.

Let I (resp. J, K) be the line joining the points $D \cap D'$ and $E \cap E'$ (resp. $D \cap D''$ and $E \cap E''$, $D' \cap D''$ and $E' \cap E''$).

Then I, J, K are concurrent.

Assume K commutative.

Let D, D' be two lines, $a, b, c \in D$ and $a', b', c' \in D'$ points.

Let i (resp. j, k) be the intersection point of the lines ab' and ba' (resp. ac' and ca', bc' and cb').

Then i, j, k are collinear.

Assume K commutative.

Let a, a' be two points, $A, B, C \ni a$ and $A', B', C' \ni a'$ lines.

Let I (resp. J, K) be the line joining the points $A \cap B'$ and $B \cap A'$ (resp. $A \cap C'$ and $C \cap A'$, $B \cap C'$ and $C \cap B'$).

Then I, J, K are concurrent.

The duality between points and lines led to a controversy amongst the geometers of the early nineteenth century. Some, like Gergonne, saw at its origin the parallelism between the linear equations of points and lines; for example, joining two points and intersecting two lines involve the same manipulations on the respective equations. In modern terms, this amounts to saying that the projective spaces of points and lines have the same structure: it's the point of view presented here. Poncelet, on the other hand, based the duality on the idea of polarity with respect to a conic (cf. section 4.3), which essentially boils down to choosing an isomorphism between a vector (or projective) space onto its dual, by means of a non-degenerate quadratic form. Since Poncelet was a general and the head of the Ecole Polytechnique, and Gergonne was a mere captain in the French artillery, it was the former's point of view that prevailed, at least among their French contemporaries. Although not going as deep into the nature of things as Gergonne's, Poncelet's point of view is suitable for the solution of metric problems (section 4.4).

1.6. The Projective Space of Circles

Affine and homogeneous coordinates

Given an orthonormal frame, a circle in a real Euclidean affine plane E has an equation of the form $x^2 + y^2 + bx + cy + d = 0$. In order to take into account circles without real points, like $x^2 + y^2 + 1 = 0$, it is well to complexify E into a complex affine space $E_{\mathbf{C}}$. Also, in order to study transformations that involve points at infinity, like inversions, it is well to take the projective closure of $E_{\mathbf{C}}$, which is essentially $\mathbf{P}_2(\mathbf{C})$, and to homogenize the coefficients of the equations of circles. Thus we'll call a *circle* the set of points in $\mathbf{P}_2(\mathbf{C})$ whose homogeneous coordinates (x, y, t)

satisfy an equation of the form

$$a(x^2 + y^2) + bxt + cyt + dt^2 = 0, \tag{24}$$

where $(a, b, c, d) \neq (0, 0, 0, 0)$. Two equations of this form define the same set if and only if they are proportional. More generally:

Theorem 10. *Two quadratic homogeneous equations $F(x, y, t) = 0$ and $G(x, y, t) = 0$ define the same subset C of a projective plane over an algebraically closed field if and only if they are proportional.*

Proof. If F is the square of a linear form, C is a line ("counted twice"), and this determines the linear form up to a scalar. Otherwise C contains three non-collinear points, which we can take as base points of P. The equations of F and G are then of the form

$$uxy + vyt + wtx = 0,$$
$$u'xy + v'yt + w'tx = 0.$$

If at least one of u, v, w vanishes, C is the union of two lines (the equation of F factors), and the assertion follows. Otherwise, for all points of C "at finite distance" ($t \neq 0$), we have the affine relations $y = -wx/(ux + v) = -w'x(u'x + v')$; this implies that $u'/u = v'/v = w'/w$, because

$$(wu' - uv')x^2 + (wv' - vw')x = 0$$

for infinitely many values of x. □

We will see in chapter 3 under what conditions this conclusion still holds over an arbitrary field. Literally speaking the conclusion doesn't hold for equations of degree ≥ 3, since $x^2y = 0$ and $xy^2 = 0$ define the same union of two lines; but see also section 8.

Thus a circle is characterized by the point of homogeneous coordinates (a, b, c, d) in a projective space $\mathbf{P}_3$, so that circles form a three-dimensional projective space. In this statement we have included, under the name of *circles*,

- true circles, for which $a \neq 0$;
- unions of the line at infinity with some other line, for which $a = 0$ and $(b, c) \neq (0, 0)$;
- the line at infinity counted twice, when $(a, b, c) = (0, 0, 0)$ and $d \neq 0$.

A *conic* is the set of points in a projective plane whose points satisfy an equation of degree two.

Theorem 11. *A conic C of $\mathbf{P}_2(\mathbf{C})$ is a circle if and only if it contains the points (x, y, t) such that $x^2 + y^2 = t = 0$, i.e., the points with homogeneous coordinates $(1, i, 0)$ and $(1, -i, 0)$.*

Proof. Let $ax^2 + bxy + cy^2 + dxt + eyt + ft^2 = 0$ the equation of C. Its intersection with the line at infinity $t = 0$ is defined by the equation $ax^2 + bxy + cy^2 = 0$; the roots (in y/x) of this equation are i and $-i$ if and only if $b = 0$ and $a = c$. □

The points $(1, i, 0)$ and $(1, -i, 0)$, common to all circles, are called *cyclic points*. Their characterization shows that they are independent of the orthonormal frame chosen in the affine plane. Lines whose point at infinity is a cyclic point, and vectors parallel to such lines, are called *isotropic* (for vectors this is the same as isotropy in the sense of quadratic forms, since the form $x^2 + y^2$ vanishes on such vectors). The isotropic lines stemming from the center of a true circle are tangent to the circle at the cyclic points; we can say they're the circle's asymptotes.

Inversions

Given a point A in a real Euclidean plane E and a real number $k \neq 0$, the *inversion of pole A and power k* is the map which associates to every point $M \neq A$ of the plane the point $M' \in D_{AM}$ such that $\overline{AM} \cdot \overline{AM'} = k$. Clearly an inversion of pole A is an involution of $P \setminus A$.

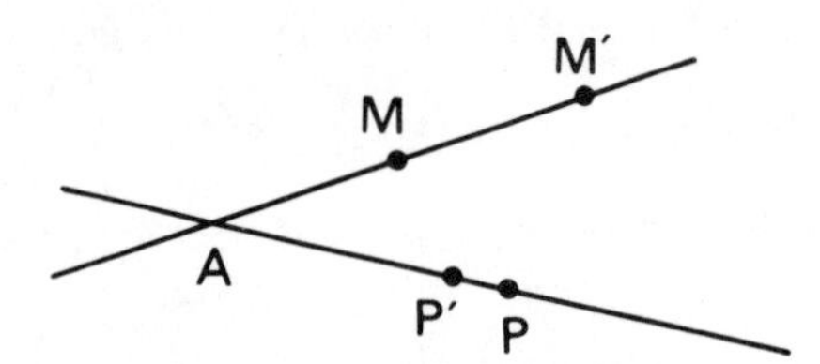

The explicit expression of $\overrightarrow{AM'}$ is $\dfrac{k}{\|\overrightarrow{AM}\|^2}\overrightarrow{AM}$. Thus, with respect to an orthonormal frame with origin A, the inversion is given by the formulas

$$x' = \frac{kx}{x^2 + y^2}, \qquad y' = \frac{ky}{x^2 + y^2}. \tag{25}$$

We extend this map to $\mathbf{P}_2(\mathbf{C})$ by homogenizing x' and y', associating to a point with homogeneous coordinates (x, y, t), whenever possible, the point with homogeneous coordinates (x', y', t') given by

$$x' = kxt, \qquad y' = kyt, \qquad t' = x^2 + y^2. \tag{26}$$

The map is now defined on the complement of the pole A and of the cyclic points I, J. Every other point of the line at infinity D_{IJ} maps to A, every point of $D_{AI} \setminus \{A, I\}$ maps to I and every point of $D_{AJ} \setminus \{A, J\}$ maps to J. On the complement of these three lines the inversion is an involution. Its fixed points are the points on the circle $x^2 + y^2 = k$; the existence of real fixed points is equivalent to $k > 0$. Since this circle characterizes the inversion, we sometimes talk about the the *inversion with respect to a given circle*.

Let us study how a circle behaves under inversion. By plugging into (24) the values of x', y', t' (instead of (x, y, t)) given by (26), we obtain

$$ak^2t^2(x^2 + y^2) + (bkxt + ckyt)(x^2 + y^2) + d(x^2 + y^2)^2 = 0.$$

We eliminate the makeweight factor $x^2 + y^2$ (which corresponds to the points on the isotropic lines, whose images are the cyclic points), to obtain the equation

$$d(x^2 + y^2) + bkxt + ckyt + ak^2t^2 = 0. \tag{27}$$

Theorem 12. *An inversion with pole at the origin and power k transforms the circle with homogeneous coordinates (a, b, c, d) into the circle with homogeneous coordinates (d, kb, kc, k^2a).* □

Thus inversion gives rise to an involutive projective transformation of the space of circles. An analysis of (24) and (27) shows that:

- a true circle not going through A ($a \neq 0$, $d \neq 0$) maps to a true circle not going through A;
- a true circle going through A ($a \neq 0$, $d = 0$) maps to the union of the line at infinity with another line, and conversely;
- the line at infinity counted twice ($a = b = c = 0$, $d \neq 0$) maps to the union $x^2 + y^2 = 0$ of the isotropic lines of A, which is the "circle of radius zero" centered at A, and conversely.

The second line corresponding to a circle C through A is orthogonal to the line that joins A with the center of C. The calculation to show this is easy, but easier yet is to observe that the figure is symmetric with respect to that line.

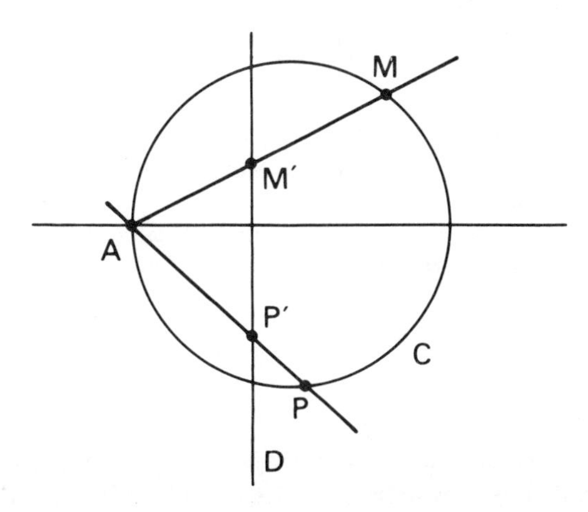

Orthogonality

A circle whose affine equation is $x^2 + y^2 - 2ux - 2vy + w = 0$ has center (u, v) and radius $(u^2 + v^2 - w)^{1/2}$. Saying that this circle is orthogonal to another circle of equation $x^2 + y^2 - 2u'x - 2v'y + w' = 0$ is saying that the square of the distance between the centers is the sum of the squares of the radii, that is,

$$(u - u')^2 + (v - v')^2 = u^2 + v^2 - w + u'^2 + v'^2 - w',$$

or again that $2(uu' + vv') - (w + w') = 0$. Thus, for two circles whose homogeneous coordinates in the space of circles are (a, b, c, d) and (a', b', c', d'), orthogonality is expressed by

$$bb' + cc' - 2(ad' + da') = 0 \tag{28}$$

(take $u = -b/2a$, $v = -c/2a$, $w = d/a$, etc., and multiply by aa'.)

Turning to the degenerate cases, (28) has the following consequences:

(1) a circle of radius zero C and an arbitrary circle $C' \neq 2D_0$ (where D_0 denotes the line at infinity) are orthogonal if and only if the center of C is on C';

(2) a true circle C and $D + D_0$ are orthogonal if and only D goes through the center of C;

(3) $D + D_0$ and $D' + D_0$ are orthogonal if and only D and D' are;

(4) C is orthogonal to $2D_0$ if and only if C is of the form $D + D_0$.

All of this, and especially (2) and (3), is quite reasonable. One can say that the center of a "circle" of the form $D + D_0$ is at infinity, in the direction orthogonal to D.

The left-hand side of (28) is a symmetric bilinear form. Its associated quadratic form is $b^2 + c^2 - 4ad$, which is non-degenerate. Its isotropic vectors, where this form vanishes, correspond to circles of radius zero ($u^2 + v^2 - w = 0$ in our original notation).

Remark. If one is only interested in true circles and their real points, one can associate to the circle of equation $x^2 + y^2 - 2ux - 2vy + w = 0$ the point $(u, v, w) \in \mathbf{R}^3$. Its projection (u, v) is the center of the circle. Circles of radius zero correspond to points on the paraboloid of revolution $w = u^2 + v^2$, circles without real points to points inside the paraboloid and ordinary circles to those outisde. Orthogonality between circles is given by polarity with respect to the paraboloid (see chapter 4).

Pencils of circles

A pencil of circles is a line in the projective space $\mathbf{P}_3(\mathbf{C})$ of circles. A (two-dimensional) linear system of circles is a plane in the same space.

If $F(x,y,t)=0$ and $G(x,y,t)=0$ are equations of two (distinct) circles in a pencil, the general equation of the circles in the pencil is

$$uF(x,y,t)+vG(x,y,t)=0, \tag{29}$$

where $(u,v)\neq(0,0)$ and two proportional pairs (u,v) yield the same circle. The points common to $F=0$ and $G=0$ are common to all the circles in the pencil; we call them the *base points* of the pencil. The cyclic points I and J are base points of all pencils.

Theorem 13. *Every point (x_0,y_0,t_0) that is not a base point of a pencil of circles is contained in exactly one circle in the pencil.*

Proof. If $F(x_0,y_0,t_0)\neq 0$, for example, take $u=G(x_0,y_0,t_0)$ and $v=-F(x_0,y_0,t_0)$ in (29). □

The bilinearity of the orthogonality relation (28) immediately shows that, if a circle is orthogonal to two circles in a pencil, it is orthogonal to all circles in the pencil. More precisely, the classical results on the orthogonality of vector subspaces with respect to a non-degenerate bilinear form take the following aspect:

Theorem 14.

(a) *The set of circles orthogonal to all the circles of a pencil $\mathcal{F}$ is a pencil $\mathcal{F}'$, called the pencil orthogonal to $\mathcal{F}$. The pencil orthogonal to $\mathcal{F}'$ is $\mathcal{F}$.*
(b) *The set of circles orthogonal to a given circle forms a two-dimensional linear system of circles. Conversely, every such system is obtained in this way.* □

For a circle of the form $D+D_0$, this linear system consists of the circles with centers on D (plus lines orthogonal to D); for a circle C of radius zero it consists of the circles that go through the center of C.

We now study the classification of pencils of circles. If a pencil $\mathcal{F}$ contains two degenerate circles, t factors out in equation (29); all points at infinity are base points and the rest of the pencil is a pencil of (concurrent or parallel) lines.

Otherwise, exactly one linear combination in (29) makes the coefficient of x^2+y^2 vanish, and $\mathcal{F}$ contains a unique degenerate circle $D+D_0$; the line D is called the *radical axis* of the pencil. Thus a pencil of this type is determined by its radical axis and one true circle.

The circle with homogeneous coordinates (a,b,c,d) has as its center the point with homogeneous coordinates $(b,c,-2a)$; this triple is a linear function of the quadruple (a,b,c,d). This means that if C runs through a pencil $\mathcal{F}$, its center draws a line L, or else is fixed. If it is fixed, we're dealing with a pencil of concentric circles, whose general equation is

$$u\big((x-x_0t)^2+(y-y_0t)^2\big)+vt^2=0$$

and whose radical axis is the line at infinity.

Otherwise L is called the *central line* of the pencil. It is orthogonal to the radical axis D because the "center" of $D + D_0$ is at infinity, in the direction perpendicular to D. In order to classify pencils of circles with real coefficients, we take L and D to be the x- and y-axes. After this normalization a pencil $\mathcal{F}$ is determined by one of its circles, say the one centered at the origin, which may, however, be imaginary. Together with D, this circle gives for $\mathcal{F}$ the affine equation

$$x^2 + y^2 - k - 2wx = 0. \tag{30}$$

The center of a circle C_w with parameter value w is $(w, 0)$, and the square of its radius is $k + w^2$. Three cases are possible, depending on the sign of k:

- $k > 0$: the circle C_0 is real, of radius $\sqrt{k}$. It intersects the radical axis at two points A and B, of coordinates $(0, \sqrt{k})$ and $(0, -\sqrt{k})$, which are the (non-cyclic) base points of the pencil. The pencil $\mathcal{F}$ is formed of all circles that contain A and B (it's easy to see that (30) is the general equation of such circles); we say that $\mathcal{F}$ is a *pencil with base points*. Such a pencil has no circle with real center and zero radius.
- $k < 0$: the circle C_0 is imaginary. The base points of the pencil are complex conjugate. The power of the origin with respect to the circles of the pencil is the positive constant $-k$, so that the origin lies outside the circles. The circle C_w has real points if and only if $w \geq \sqrt{-k}$. The circles of radius zero and centered at $(\sqrt{-k}, 0)$ and $(-\sqrt{-k}, 0)$ belong to $\mathcal{F}$; they are called the *Poncelet points* of $\mathcal{F}$, and $\mathcal{F}$ is said to be a *pencil with Poncelet points*.
- $k = 0$: in this case (30) is the general equation of the circles tangent to the radical y-axis at the origin. The only non-cyclic base point is the origin, counted twice (sometimes one counts the origin together with an "infinitely close point in the y-direction"). If $\mathcal{F}$ is of this type we say that $\mathcal{F}$ is a *pencil of tangent circles*.

Let's study the pencils orthogonal to those of each of the types above. If $\mathcal{F}$ is degenerate things are pretty easy:

Concurrent lines going through A	Concentric circles centered at A
Parallel lines	Parallel lines orthogonal to the first

If $\mathcal{F}$ is given by equation (30), the central line (the x-axis) and the circle $x^2 + y^2 + k = 0$ are orthogonal to every circle in $\mathcal{F}$; thus the orthogonal pencil has general equation

$$x^2 + y^2 + k - 2w'y = 0. \tag{31}$$

The central line and the radical axis have exchanged roles. Since k has become $-k$, the three types above are interchanged:

Pencil with base points	Pencil with Poncelet points

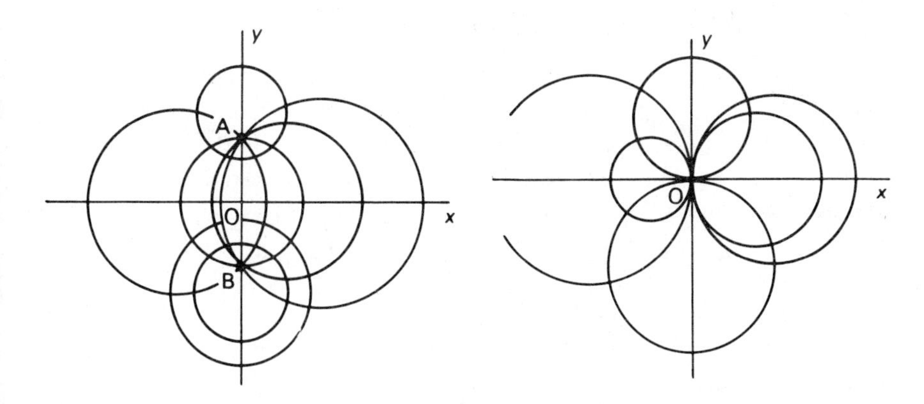

Pencil with Poncelet points
Pencil of tangent circles

Pencil with base points
Pencil of tangent circles

The base points of one pencil are the Poncelet points of its orthogonal.

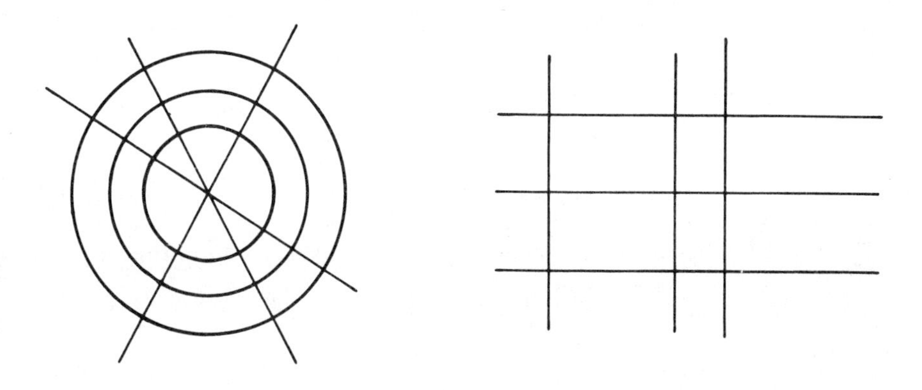

Finally, since inversions are projective transformations of the space of circles, they transform pencils into pencils, and this allows us to simplify the study of pencils by reducing them to "standard forms". Orthogonality (28) is preserved under inversion (theorem 12), so orthogonal pencils are taken into orthogonal pencils. If $\mathcal{F}$ is a pencil with base points A and B, an inversion j with pole A takes $\mathcal{F}$ into the pencil of lines going through $j(B)$. By orthogonality, a pencil with Poncelet points A and B is taken under j into the pencil of concentric circles centered at $j(B)$. Finally, the pencil of circles tangent to some line D at the origin is taken, under an inversion whose pole is the origin, into the pencil of lines parallel to D.

1.7. The Projective Space of Conics

Irreducibility

Let the field of scalars K be commutative. A *conic* is the set of points on a projective plane P whose homogeneous coordinates (x, y, t) satisfy a homogeneous quadratic equation:

$$F(x,y,t) = ax^2 + by^2 + ct^2 + a'yt + b'xt + c'xy = 0. \tag{32}$$

The set of points of the conic determines the quadratic form F up to a scalar when K is algebraically closed (theorem 10) and in many other cases (section 3.1).

A conic is said to be *irreducible* or *non-degenerate* if F is irreducible over the algebraic closure of K. Otherwise F is the product of two linear forms, so the conic is either the union of two distinct lines, or a single line "counted twice" (if F is proportional to a square).

Theorem 15. *A conic C is non-degenerate if and only if all of its points are simple.*

Proof. If C splits into two lines, distinct or not, a point in the intersection of the two lines is double (section 3). Conversely, if C has a multiple point A, the line that joins it with another point of C has at least three points in common with C, so it is entirely contained in C; the rest of C is a line as well. □

Thus C is non-degenerate if and only if the equations $F'_x = F'_y = F'_t = 0$ and $F = 0$ have a common non-trivial solution. This implies that the determinant of the three derivatives, which are linear, is zero, that is

$$\begin{vmatrix} 2a & c' & b' \\ c' & 2b & a' \\ b' & a' & 2c \end{vmatrix} = 0. \tag{33}$$

By Euler's formula, this condition is enough in characteristic $\neq 2$; incidentally, it expresses that the quadratic form F is degenerate (that is, of rank 1 or 2).

In characteristic 2, the three derivatives F'_x, F'_y and F'_t automatically vanish for $(x, y, t) = (a', b', c')$, and one sees without much trouble those are the only values of (x, y, t) (up to a scalar factor) for which this happens, unless $a' = b' = c' = 0$, in which case F is a square over the algebraic closure of K. Thus C is degenerate if and only if $F(a', b', c') = 0$, that is,

$$aa'^2 + bb'^2 + cc'^2 + a'b'c' = 0.$$

The point (a', b', c') is equally interesting when the conic C (still in characteristic 2) is irreducible: then all the tangents to C go through this point! To see this, recall from section 3 the formula (23) for the tangent to an algebraic curve:

$$\begin{aligned} a'F'_x + b'F'_y + c'F'_t &= a'(c'y + b't) + b'(c'x + a't) + c'(b'x + a'y) \\ &= 2a'c'y + 2b'c'x + 2b'a't = 0. \end{aligned}$$

Intersection of two conics

Theorem 16. *Let C and C' be conics not having a line in common. There are at most four points common to C and C', and exactly four if K is algebraically closed and we count multiplicities.*

Proof. This is clear if at least one of the conics is degenerate. Otherwise, take $A, B \in C$ such that $A \notin C'$ and $A \neq B$. Choose a projective coordinate system in which A is the point $(0,1,0)$, the tangent at A is the line at infinity, B is the point $(0,0,1)$ and the intersection of the line at infinity with the tangent at B is $(1,0,0)$. Write the equation of C in the form (32).

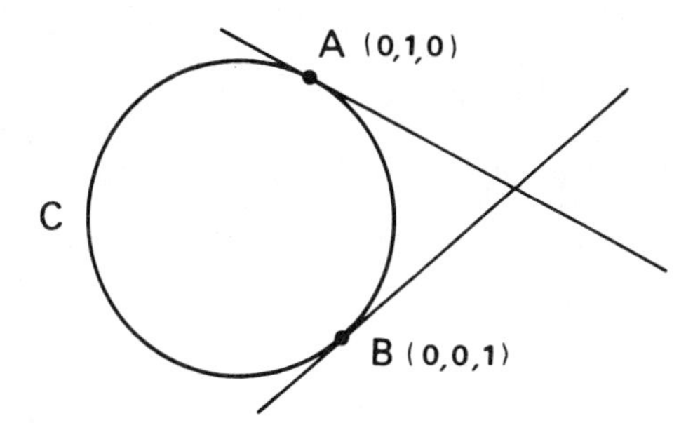

Intersecting (32) with the line at infinity $t = 0$ we see that $b = c' = 0$; similarly, observing that $y = 0$ is tangent to C at $(0,0,1)$, we see that $c = b' = 0$. The equation of C then reduces to $ax^2 + a'yt = 0$. Since C and C' have no common point at infinity, we can work in affine coordinates, where, replacing y by $-a'y/a$, the equation of C is $y = x^2$ (the most simple-minded of parabolas!). Plugging this into the equation of C', we obtain an equation in x that is exactly of degree four, since the term uy^2 of the equation of C' is non-zero (recall that $(0,1,0)$ is not on C').

In elementary geometry one learns that two circles have at most two common points (of multiplicity one). The other two are the cyclic points (section 6); but they're invisible, being at infinity and imaginary, to boot.

Theorem 17. *Five distinct points in a projective plane, no four of which are collinear, uniquely determine a conic.*

Proof. The requirement that each point (x_i, y_i, t_i) $(i = 1, 2, \ldots, 5)$ is contained in the conic is expressed by five homogeneous linear equations in the six coefficients a, b, c, a', b', c' of a general conic (32). These equations have a non-trivial solution but it is not *a priori* clear that the solution is unique up to a scalar (that is, that the equations are independent). Suppose there are two distinct conics C and C' containing the five points. By theorem 16, they have a line D in common. By assumption, D contains at most three of the five points; the others, of which there are at least two, are enough to determine the second line of C and of C'. This shows that $C = C'$. □

If four of the five points lie on a line D, the union of D with any line going through the fifth is a solution to the problem.

Linear systems of conics

If we call the six coefficients (a, b, c, a', b', c') of a conic (equation (32)) its homogeneous coordinates, we see that conics form a five-dimensional projective space. The projective linear subspaces of this space are called *linear systems of conics*, or *pencils* when one-dimensional. Circles form a three-dimensional linear system of conics, containing all conics that go through the cyclic points I and J.

In order to study the conics that go through two fixed points of a projective plane, it may be convenient to apply a projective transformation and map them to the cyclic points. The problem is then reduced to the study of circles.

A pencil of conics has a general equation of the form

$$uF(x, y, t) + vG(x, y, t) = 0, \tag{33}$$

where F and G are non-proportional homogeneous degree-two polynomials, u and v are no both zero, and two proportional pairs (u, v) generate the same conic. A pencil is uniquely determined by two conics belonging to it. The set where $F(x, y, t) = G(x, y, t) = 0$ belongs to all conics in the pencil; we call it the *base* of the pencil. As in the case of circles (theorem 13) we see that every point that is not in the base is contained in exactly one conic in the pencil.

We will rule out the case when the base of the pencil contains a line D, because in this case the linear form defining D is a factor in (33), and every conic in the pencil is the union of D with some other line D' that runs through a pencil of lines. Apart from this case, every pencil of conics has at most four base points (theorem 16).

The most general case is when a pencil has four distinct base points p, q, r, s. No three of them can be on the same line D, for if they were, D would have three common points with every conic in the pencil, hence would be contained in them, a case that has been ruled out. Thus the degenerate conics of the pencil can only be $D_{pq} + D_{rs}$, $D_{pr} + D_{qs}$ and $D_{ps} + D_{qr}$. If we take p, q, r, s, in this order, as elements of a projective frame, the general equation of the pencil reduces to the form

$$ux(y - t) + vy(x - t) = 0. \tag{34}$$

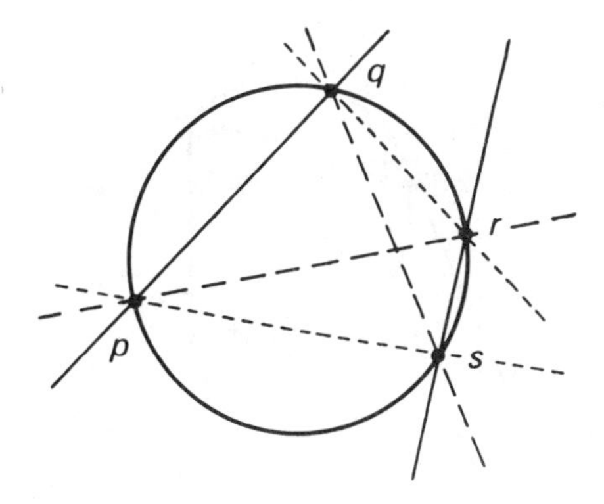

Conversely, we have the following result:

Theorem 18. *The set of conics containing four points no three of which are collinear is a pencil.*

Proof. Consider a projective frame formed by the four points. The conics containing the three vertices have equations of the form $axy+byt+czt=0$, so they form a two-dimensional linear system. The condition of passing through $(1,1,1)$ is that $a+b+c=0$, which gives us (34) with $u=-c$ and $v=-b$. □

For another proof, take two conics $F=0$ and $G=0$ containing the four points (two degenerate conics, for instance). Any conic C that goes through the four points belongs to the pencil $\mathcal{F}$ of equation $uF+vG=0$, for if m is a fifth point on C, the unique conic $C' \in \mathcal{F}$ containing m (cf. section 6) coincides with C, the two having five points in common (theorem 16).

If a pencil has three base points p,q,r, they are non-collinear (otherwise we're in the excluded case). At one of the points, say p, any two conics of the pencil must have an intersection with multiplicity two; since the only conic of the pencil having p as a double point is $D_{pq}+D_{pr}$, the other conics are tangent to one another at p, and thus all tangent to the same line $T \ni p$. There is only one more degenerate conic in the pencil, namely $T+D_{qr}$. If we choose a projective frame so that $p=(0,0,1)$, $q=(0,1,0)$, $r=(0,0,1)$ and $(1,1,1) \in T$, the general equation of the pencil takes the reduced form

$$uxy+vt(y-x)=0.$$

Example. A pencil of tangent circles, where p is the tangency point and q,r are the cyclic points.

If a pencil has two base points p,q, two cases can occur: either the conics intersect one another with multiplicity two at both p and q, or they

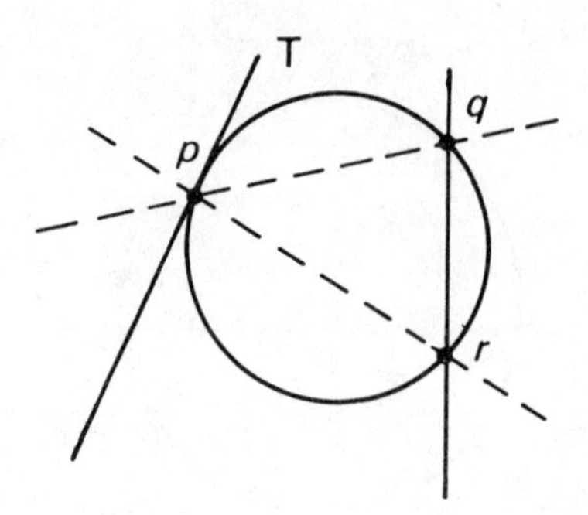

intersect with multiplicity one at p and three at q. Consider the first case. The pencil has a unique conic with a double point at p, otherwise we're in the excluded case; and this conic must be of the form $2D_{pq}$ or $D_{pq} + D'$. Similarly for q. Now the pencil cannot afford both $D_{pq} + D'$ and $D_{pq} + D''$, with $D' \ni p$ and $D'' \ni q$, because that would create a third base point, $D \cap D''$. Thus $2D_{pq}$ is the only conic to have double points at p or q. All others conics are tangent at p to a fixed line T, and at q to a fixed line U; we're dealing with a pencil of *bitangent conics*. The only other degenerate conic is $T + U$. If we choose a projective frame consisting of p, q, a point in T and a point in U, in this order, the general equation of the pencil takes the reduced form

$$uxy + vt^2 = 0. \tag{35}$$

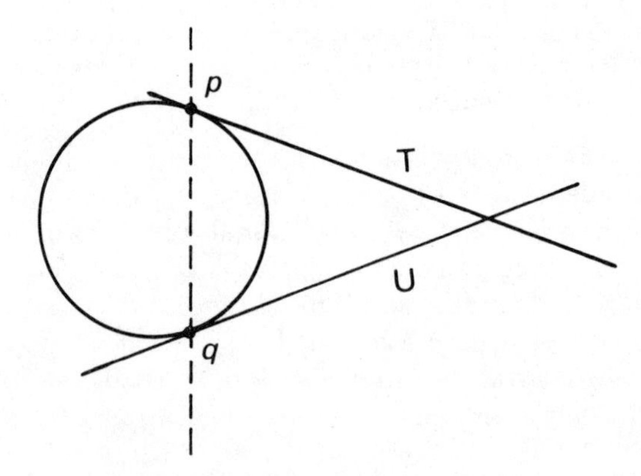

Example. A pencil of concentric circles, since they're all tangent at the cyclic points to the isotropic lines emanating from their center.

Now let's take the case of multiplicites 1 and 3. Since a conic cannot have triple points, the only degenerate conic in the pencil must be $D_{pq} + T$, where T is the common tangent at p to the conics in the pencil. We have

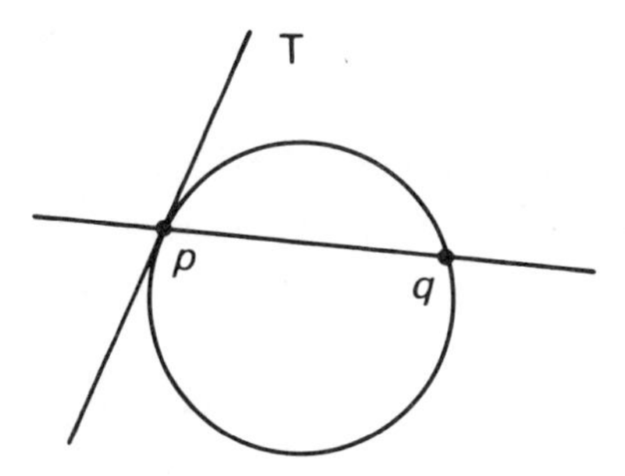

to express the condition that they intersect with multiplicity three at p; this is an affine question.

Let p be the origin, q the point at infinity on the x-axis and T the y-axis. The affine equation of a conic tangent to T at p and going through q can be written $x + cxy + dy^2 = 0$, or $(1 + cy)x + dy^2 = 0$. The common points to this conic and to another one, say $(1 + c'y)x + d'y^2 = 0$, satisfy $(1 + cy)d'y^2 - (1 + c'y)dy^2 = 0$. The (double) solution $y = 0$ gives $x = 0$, which is p. There remains $(dc' - cd')y = d' - d$. But $cd' - dc' \neq 0$, because otherwise the two conics would have the same points at infinity, for a total of three base points. If $d - d' \neq 0$, we'd have a solution with $y \neq 0$ at finite distance, hence distinct from p; this is impossible. Thus $d' = d$ and all the conics in the pencil have the same y^2 coefficient in $x + cxy + dy^2 = 0$. Replacing x by x/d and homogenizing, we get the general equation

$$u(xt + y^2) + vxy = 0. \tag{36}$$

Finally, let's examine the poor pencils who only have one base point p. The conics of such a pencil have an intersection of multiplicity four at p. Those that are degenerate are necessarily the unions of two lines going through p. If the pencil contains two such unions, say $F(x, y) = 0$ and $G(x, y) = 0$, where we have assumed p to be the origin and F and G are homogeneous quadratic polynomials, the general equation of the pencil is

$$uF(x, y) + vG(x, y) = 0 \tag{37}$$

and all its elements degenerate into two lines going through p.

If the pencil contains two double lines, its equation becomes $ux^2 + vy^2 = 0$. Then there are no other double lines, except in characteristic 2, where all conics are double lines.

Now, if the pencil contains at most one degenerate conic, two non-degenerate conics intersect with multiplicity four at p (we say that they are *superosculating*), and in particular they're both tangent at p to a line T. By making p the origin and T the y-axis the affine equations of two conics in the pencil can be written $x + F(x, y) = 0$ and $x + G(x, y) = 0$, where F and G are homogeneous polynomials of degree two. By subtraction the conic $G(x, y) - F(x, y) = 0$ is in the pencil, and it is necessarily degenerate, hence of the form $2T$; thus $G = F + kx^2$. The general equation of the pencil is then

$$u(xt + F(x,y)) + vx^2 = 0. \tag{38}$$

One cannot normalize the quadratic form $F(x, y)$ because the points at infinity of the conics in the pencil vary.

To summarize, the only pencils containing only degenerate conics are:

- those whose elements are unions of a fixed line and a line that rotates around a point; and
- those whose elements are unions of two lines going through a fixed point.

The only pencils whose elements have multiple points that vary are pencils of double lines of the form $ux^2 + yv^2 = 0$ in characteristic 2.

All other pencils have at most three degenerate elements, because saying that a conic is degenerate is saying that a certain polynomial of degree three in its coefficients vanishes (equation (33) and following note concerning characteristic 2). Thus the condition for a conic $uF(x, y, t)+vG(x, y, t) = 0$ of a pencil to be degenerate is expressed by a homogeneous cubic equation $H(u, v) = 0$ (and $H = 0$ if all the conics in the pencil are degenerate).

This yields a geometric explanation for the fact that the problem of solving of a quartic equation can be reduced to solving cubics and quadratics. Leaving aside the case of characteristic two for the sake of simplicity, a quartic can be reduced to the form $x^4 + ax^2 + bx + c = 0$. If we set $y = x^2$ this is equivalent to the system $y - x^2 = 0$, $y^2 + ay + bx + c = 0$, which is the intersection of two pararabolas. These two parabolas belong to the pencil whose general affine equation is

$$y^2 + ay + bx + c + v(y - x^2) = 0.$$

A short calculation applied to (33) shows that the parameter values of degenerate conics in this pencil are the roots of $v(v + a)^2 + 2cv + b^2 = 0$. Let v_0 be such a parameter value. The double point of the corresponding degenerate conic is a rational function of v_0, because its coordinates are the solutions of a linear system. The slopes of the two lines that make up the conic are solutions of a quadratic equation. Finally, their intersections with the parabola $y = x^2$, that is, the base points of the pencil, are also given by quadratic equations.

Since equations of degree two and three are soluble by radicals (the latter by Cardano's formula, for instance), the same is true about quartics.

1.8. Projective Spaces of Divisors in Algebraic Geometry

We have seen in section 3 that is was worthwhile to count intersection points with their multiplicities. We have also called conics whose equation is a square "double lines". And we have noticed that two non-proportional equations, like $x^2y = 0$ and $xy^2 = 0$, can define the same algebraic subset.

We say that a (projective or affine) hypersurface is *irreducible* if its defining polynomial is irreducible. And we define a *divisor* in an affine or projective space to be a formal linear combination, with integer coefficients, of irreducible hypersurfaces. Thus a divisor can be written $n_1H_1 + \cdots + n_qH_q$.

This idea has already crept in, when we wrote $D + D'$ and $2D$.

Now let K be an algebraically closed field and $F(x_0, \ldots, x_n)$ a homogeneous polynomial of degree d over K. Since the ring of polynomials in $n+1$ variables over K is a unique factorization domain, F can be decomposed in an essentially unique way into a product $F = F_1^{n_1}F_2^{n_2}\ldots F_q^{n_q}$ of (necessarily homogeneous) irreducible polynomials, where the F_j are pairwise non-proportional. If we denote by H_j the hypersurface $F_j(x_0, \ldots, x_n) = 0$, we define the *divisor* of F as

$$(F) = n_1H_1 + \cdots + n_qH_q. \tag{39}$$

Theorem 19. *Two homogeneous polynomials F and G have the same divisor if and only if they are proportional.*

Proof. Sufficiency is obvious. To prove necessity one must show that, if two irreducible homogeneous polynomials P and Q define the same hypersurface, they are proportional. Now an elementary theorem of algebraic geometry, the Hilbert Nullstellensatz, says that (over an algebraically closed field) if a polynomial Q vanishes on every zero common to polynomials $P_1, \ldots, P_r$, there is a power Q^s of Q in the ideal generated by $P_1, \ldots, P_r$ in the ring of polynomials. We will not prove this theorem, but its use is straightforward: if P and Q define the same hypersurface, some power Q^s is a multiple of P, and in fact Q itself divides P because of unique factorization and the irreducibility of Q. Similarly, P divides Q. Thus Q/P is a unit in the ring of polynomials, hence a non-zero constant.

Notice that, by definition, every divisor of $\mathbf{P}_n(K)$ is the divisor of a homogeneous polynomial, determined up to a scalar factor; we let the *degree of this divisor* be the degree of this homogeneous polynomial. Since homogeneous polynomials of degree d in $n+1$ variables form a vector space over K, of dimension $\binom{n+d}{d}$, we see that divisors of degree d form a projective space, of dimension $\binom{n+d}{d} - 1$.

Thus, over $\mathbf{P}_2$ and $\mathbf{P}_3$, divisors of degree d form a projective space of dimension $\frac{1}{2}(d+2)(d+1) - 1 = \frac{1}{2}d(d+3)$ and $\frac{1}{6}d(d^2 + 6d + 11)$, respectively. The divisors of $\mathbf{P}_1$ are formal linear combinations of points; in degree d, they form a d-dimensional projective space. The homogeneous coordinates of such a divisor are the coefficients of the corresponding (degree-d) homogeneous polynomial $F(u, v)$, which in are elementary symmetric functions of the homogeneous coordinates of the points on the divisor. For example, the sum $P + P'$ of points (x, y) and (x', y') has homogeneous coordinates $(xx', -(xy' + yx'), yy')$ (take the product of the linear forms $vx - uy$ and $vx' - uy'$).

CHAPTER 2

One-Dimensional Projective Geometry

2.1. Cross-ratios and Rational Maps

Throughout this chapter we assume that the field of scalars is commutative. Recall that a *projective line* is a one-dimensional projective space, that a projective frame on a projective line D is made up of three distinct points (a, b, c) (section 1.1) and that there exists a unique projective transformation taking a projective frame into another (1.2); in particular, the projective group of a projective line D acts simply transitively on the set of triples of distinct points of D.

The standard projective line

Given a projective line D and a projective frame for it, each point in D corresponds to a homogeneous class, with all its proportional pairs of homogeneous coordinates (x, y). When $x \neq 0$, these pairs are uniquely determined by the ratio $t = y/x$; on the other hand, when $x = 0$, the coordinate class is also uniquely determined, and we write $t = \infty$, where ∞ is a symbol not in K. Thus we have a correspondence between points in D and their values $t \in K \cup \{\infty\}$. This correspondence is bijective, and canonically so when $D = \mathbf{P}_1(K)$. The set $\hat{K} = K \cup \{\infty\}$, with the projective structure transferred from $\mathbf{P}_1(K)$, is called the *standard projective line*, and the frame $\{\infty, 0, 1\}$ on it is the *standard projective frame*. The element t associated with a point of D is sometimes called its *projective coordinate ratio* (in the coordinate system being considered).

When K has a non-discrete absolute value f, we give $\hat{K}$ the topology whose basis consists of all open subsets of K (in the topology induced by f) together with all sets of the form $C \cup \{\infty\}$, where C is the complement of a ball in K. When K is locally compact, $\hat{K}$ is the Alexandroff compactification of K. For $K = \mathbf{R}$ (resp. $K = \mathbf{C}$) this set is the circle (resp. the sphere).

Let $h : D \to D'$ be a projective transformation from a projective line onto another. If D and D' are given projective frames, the homogeneous coordinates (x', y') of the image $h(m)$ of a point m of D are linear functions of the homogeneous coordinates (x, y) of m (section 1.2):

$$x' = dx + cy, \quad y' = bx + ay \qquad (ad - bc \neq 0).$$

Thus the projective coordinate ratio $t' = y'/x'$ of $h(m)$ is given, as a function of $t = y/x$, by the formula

$$t' = (at + b)/(ct + d), \tag{40}$$

which holds, *a priori,* for t finite and $ct + d \neq 0$. If $ct + d = 0$, we have $x' = dx + cy = 0$, whence $t' = \infty$, which leads us to assign the value ∞ to a fraction whose denominator vanishes. For $t = \infty$, we have $x = 0$, whence $y'/x' = a/c$ if $c \neq 0$ and $t' = y'/x' = \infty$ if $c = 0$. With these conventions, formula (40) defines a bijective map from (all of) $\hat{K}$ onto itself, which is the translation of the projective transformation h in terms of projective coordinate ratios. Conversely, any map $\hat{K} \to \hat{K}$ of the form $t' = (at + b)/(ct + d)$, for $ad - bc \neq 0$, can be interpreted as a projective transformation.

If K has an absolute value and $\hat{K}$ has the topology above, every projective transformation on $\hat{K}$ is continuous, since it extends by continuity the ordinary function $f(t) = (at + b)/(ct + d)$ defined on $K \setminus \{-d/c\}$ (or on K, if $c = 0$).

Cross-ratios

Definition 5. Let a, b, c, d be points on a projective line D, the first three of which are distinct. The *cross-ratio* of the four points, denoted by (a, b, c, d) or $(a, b; c, d)$, is defined as $h(d) \in \hat{K}$, where h is the unique projective transformation $D \to \hat{K}$ that takes a, b and c to ∞, 0 and 1, respectively.

Theorem 20. *Let D and D' be projective lines, a, b, c, d points on D and a', b', c', d' distinct points on D'. There exists a projective transformation $u : D \to D'$ taking a, b, c, d into a', b', c', d', respectively, if and only if the cross-ratios (a, b, c, d) and (a', b', c', d') are equal.*

Proof. Let h (resp. h') be the unique projective transformation of D (resp. D') onto $\hat{K}$ that takes a, b, c (resp. a', b', c') to $\infty, 0, 1$. If u is the projective transformation from D onto D' that takes a, b, c to a', b', c', we have $h(m) = h'(u(m))$ for every $m \in D$, by uniqueness. If $u(d) = d'$, we deduce that

$$(a,b,c,d) = h(d) = h'(d') = (a',b',c',d').$$

Conversely, if the two cross-ratios are the same, we have $h(d) = h'(d')$, hence also $h(d) = h'(u(d))$; this implies $u(d) = d'$, since h' is injective. □

Theorem 21. *Let a, b, c, d be elements of $\hat{K}$, the first three of which are distinct. With the usual conventions about operations with 0 and ∞, the cross-ratio (a,b,c,d) is given by*

$$(a,b,c,d) = \frac{(c-a)(d-b)}{(c-b)(d-a)}. \tag{41}$$

For example, if $c = \infty$, the expression $(c-a)/(c-b)$ evaluates to 1; if $d = a$, we get $(a,b,c,d) = \infty$.

Proof. The projective transformation $h : \hat{K} \to \hat{K}$ that takes a, b, c into $\infty, 0, 1$ has a as a pole and b as a zero, so it must be of the form $h(t) = k(t-b)/(t-a)$. Since $h(c) = 1$, we must have $k = (c-a)/(c-b)$, which proves (41). □

Corollary. *The cross-ratio (a,b,c,d), seen as a function of a, b, c or d separately, is of the form (40), and hence a projective transformation.* □

Rational maps

A *rational fraction* on a field K is an element $r(T)$ of the field of fractions $K(T)$ of the ring $K[T]$ of polynomials in one variable. Thus $r(T)$ can be written as the quotient $r(T) = p(T)/q(T)$ of two polynomials; for simplicity, we can assume that $p(T)$ and $q(T)$ are relatively prime, in which case we say that the expression $r(T)$ is reduced. If we in addition require that the denominator $q(T)$ be a monic polynomial, $r(T)$ has a unique reduced expression.

Theorem 22. *Every rational fraction $r(T) = p(T)/q(T)$ defines a map $\hat{K} \to \hat{K}$ extending the usual evaluation map $t \mapsto p(t)/q(t)$ (which is only defined on K minus the zeros of $q(T)$).*

Proof. Assume $r(T) = p(T)/q(T)$ is in reduced form. For $t \in K$ such that $q(t) \neq 0$, take $r(t) = p(t)/q(t)$. For $t \in K$ such that $q(t) = 0$ (which implies $p(t) \neq 0$, since p and q are relatively prime) we set $r(t) = \infty$. Finally, $r(\infty)$ is calculated by setting $T = 1/U$ and $s(U) = r(1/U)$, and by forming $s(0)$, with the following results: 0 if $d^0(p) < d^0(q)$; a/b if $d^0(p) = d^0(q)$, where a and b stand for the highest-degree coefficients of p and q; and ∞ if $d^0(p) > d^0(q)$. □

If K has an absolute value and $\hat{K}$ has the topology above, the map given by theorem 22 is the extension by continuity of the ordinary rational function $t \mapsto p(t)/q(t)$.

A map as given by theorem 22 is called a *rational map* from $\hat{K}$ into itself.

Theorem 23. *If $r(T) = p(T)/q(T)$ is a non-constant rational fraction in reduced form, $K(T)$ is a finite extension of the subfield $K(r(T))$ generated by K and $r(T)$, and*

$$[K(T): K(r(T))] = \max(d^0p, d^0q). \tag{42}$$

Proof. Set $r = r(T)$. We show first that r is transcendent over K, that is, that it is not the root of a non-zero polynomial with coefficients in K. Otherwise we would have $a_n r^n + \cdots + a_1 r + a_0 = 0$ with $a_n \neq 0$ and $a_0 \neq 0$, whence $a_n p^n + a_{n-1}p^{n-1}q + \cdots + a_1 pq^{n-1} + a_0 q^n = 0$, so that p divides q^n. But p and q are relatively prime, so p must be constant; similarly, q and r are also constant, a contradiction. This shows that the subring $K[r]$ is isomorphic to the ring of polynomials over K.

Let X be another indeterminate. The polynomial $q(X) - rp(X)$ over $K(r)$ has T as a root and has degree $\max(d^0p, d^0q)$: there are no simplications because r is transcendent. There remains to show that this polynomial is irreducible over $K(r)$, and, *a fortiori*, over $K[r]$. Now $K[r][X]$ can be seen as the ring $K[r, X]$ of polynomials in two variables, which can also be written $K[X][r]$. As a polynomial in r over $K[X]$, $q(X) - rp(X)$ is irreducible over $K[X]$ (being of degree one) and primitive over $K[X]$ (since $q(X)$ and $p(X)$ are relatively prime). The irreducibility of $q(X) - rp(X)$ now follows from standard results on the decomposition of polynomials over factorial rings. □

The number $\max(d^0p, d^0q)$ is called the *degree* of the (non-constant) rational fraction $r(T)$. It is denoted by d^0r. If K is algebraically closed, the corresponding rational map from $\hat{K}$ into itself is onto, and its fibers (inverse images of points) all have the same cardinality d^0r, counting multiplicities.

For a given $a \in \hat{K}$, the fiber of a is made up of the (finite or infinite, cf. section 1.3) roots of the equation $q(X) - ap(X) = 0$.

Rational fractions of degree one are exactly those of the form $r(T) = (aT + b)/(cT + d)$, with $ad - bc \neq 0$. The rational maps derived from them are bijective, but in characteristic $p \neq 0$ they are not the only ones with that property (think of $r(T) = T^p$, for example.)

If $r(T)$ and $s(T)$ are non-constant rational fractions, we can replace T by $s(T)$ in s to form the composition $s(r(T))$, denoted by $s \circ r$. Its rational map is the composition of the rational maps of s and r.

Theorem 24. *If $r(T)$ and $s(T)$ are non-constant rational fractions, the degree of $s \circ r$ is the product of the degrees of r and s.*

Proof. Since $r(T)$ is transcendent over K (theorem 23), there is a K-isomorphism $f : K(T) \to K(r(T))$ such that $f(T) = r(T)$. This isomorphism takes $s(T)$ to $s(r(T))$, so that

$$[K(T): K(s(T))] = [K(r(T)): K(s(r(T)))].$$

The theorem now follows from theorem 23 and the multiplicativity of degrees of extension fields, applied to $K(s(r(T))) \subset K(r(T)) \subset K(T)$. □

Corollary. *If a rational map from $\hat{K}$ into itself is invertible and its inverse is a rational map, they are both projective transformations.*

Proof. The condition $d^0 r d^0 s = 1$ implies $d^0 r = 1$ and $d^0 s = 1$, so r and s are both of the form (40). □

It follows also from theorem 24 that the degree of a rational map does not change if we compose it (at the right or at the left) with a projective transformation. Since base changes on a projective line correspond to projective transformations, the notion of a *rational map from one projective line into another* is well-defined, as is the *degree* of such a map. Projective transformations are the degree-one rational maps, and degrees get multiplied under composition (theorem 24).

The translation of the corollary to theorem 24 into the language of fields is the following: every K-automorphism of the field of rational fractions $K(T)$ takes T into a rational fraction of degree 1.

2.2. Cross-ratios and permutations

Given four (distinct) points a_1, a_2, a_3, a_4 on a projective line D and a permutation $s \in \mathcal{S}_4$, can one calculate the cross-ratio $(a_{s(1)}, a_{s(2)}, a_{s(3)}, a_{s(4)})$ as a function of (a_1, a_2, a_3, a_4)? The answer is yes, because we have:

Lemma. *If $(a_1, a_2, a_3, a_4) = (a'_1, a'_2, a'_3, a'_4)$ then*

$$(a_{s(1)}, a_{s(2)}, a_{s(3)}, a_{s(4)}) = (a'_{s(1)}, a'_{s(2)}, a'_{s(3)}, a'_{s(4)}).$$

Proof. The unique projective transformation that takes a_i to a'_1 (theorem 20) takes $a_{s(i)}$ to $a'_{s(i)}$. □

Thus the cross-ratio $(a_{s(1)}, a_{s(2)}, a_{s(3)}, a_{s(4)})$ is uniquely determined by $t = (a_1, a_2, a_3, a_4)$ and the permutation s. We will denote it by $R_s(t)$. We obviously have

$$R_{ss'}(t) = R_s(R_{s'}(t)), \qquad \text{for } s, s' \in \mathcal{S}_4. \tag{43}$$

To compute $R_s(t)$ for $t \in \hat{K}$, recall that $t = (\infty, 0, 1, t)$ and apply the permutation s: for example, if s is the transposition $(2, 3)$ we have $R_s(t) =$

$(\infty, 1, 0, t) = 1 - t$, by theorem 21. Since the cross-ratio, as a function of each of its variables, is given by a rational map of degree one, we deduce the following result:

Theorem 25. *The "permuted cross-ratio" $R_s(t)$ is a rational function of degree one of t, and the map $s \mapsto R_s$ is a homomorphism of the symmetric group $\mathcal{S}_4$ into the projective group of $\hat{K}(= \mathrm{PGL}(K^2))$.* □

Here's the actual calculation of $R_s(t)$:

- By inspection of (41), we see that the cross-ratio remains unchanged under the permutations $(1,2)(3,4)$, $(1,3)(2,4)$ and $(1,4)(2,3)$. Together with the identity, these permutations form an index-six subgroup $\mathbf{H}$ of $\mathcal{S}_4$. Thus there exist at most $24/4 = 6$ distinct projective transformations R_s.
- For $s = (1,2)$, the same formula shows that $R_s(t) = 1/t$. For $s' = (2,3)$, we get $R_{s'}(t) = (\infty, 1, 0, t) = (t-1)/(0-1) = 1 - t$.
- By composing the projective transformations R_s and $R_{s'}$ above, we obtain the six projective transformations $R(t) = t, 1/t, 1-t, 1/(1-t), 1 - 1/t, t/(t-1)$.

We can summarize this in the following result:

Corollary. *The homomorphism $s \mapsto R_s$ of $\mathcal{S}_4$ into the projective group has as its image the order-six subgroup G made up of*

$$t, \quad \frac{1}{t}, \quad 1-t, \quad \frac{1}{1-t}, \quad \frac{t-1}{t}, \quad \frac{t}{t-1}. \tag{44}$$

It kernel is the index-six subgroup $\mathbf{H}$ *above, which is consequently normal.* □

Notice that the six projective transformations in (44) are distinct, even over $\mathbf{F}_2$. They act freely on the set $\{\infty, 0, 1\} \subset \hat{K}$; thus $\mathbf{G} = \mathcal{S}_4/\mathbf{H}$ is isomorphic to the symmetric group $\mathcal{S}_3$.

Since $\mathbf{F}_2$ only has the three elements $\infty, 0, 1$, the group $\mathrm{PGL}(\mathbf{F}_2^2)$ is isomorphic to $\mathcal{S}_3$.

2.3. Harmonic Division

We will now study the orbits of the action of the order-six group $\mathbf{G}$ on $\hat{K}$. They have six elements, except for the values of t that make at least two of the expressions in (44) equal. To find out what these values are, notice that we can restrict ourselves to the equations

$$t = t, \quad t = \frac{1}{t}, \quad t = 1 - t, \quad t = \frac{1}{1-t}, \quad t = \frac{t-1}{t}, \quad t = \frac{t}{t-1},$$

since we're discussing a group action. We find:

(a) $t = 1$; the other values given by (44) are then ∞ and 0.

(b) $t = -1$; the other values are 2 and $\frac{1}{2}$.

(a′) $t = \infty$; the other values are 0 and 1. We're in case (a).

(b′) $2t - 1 = 0$, hence $t = \frac{1}{2}$; the other values are -1 and 2. We're in case (b).

(c) $t(1-t) = 1$, hence $t^2 - t + 1 = 0$, and $t = -j$ or $-j^2$, where j is a cube root of unity. The other values are $t = -j^2$ or $-j$, respectively.

(c′) $t = (1-t)/t$, hence $t^2 - t + 1 = 0$. We're back in case (c).

(a″/b″) $t = t/(1-t)$, hence $t = 0$ or 2. We're back in case (a) or (b), respectively.

Theorem 26. *The orbits of* $\mathbf{G}$ *on* $\hat{K}$ *all have six elements, except for* $\{\infty, 0, 1\}$, $\{-1, 2, \frac{1}{2}\}$ *and* $\{-j, -j^2\}$. □

The cross-ratio (a, b, c, d) belongs to the first of these special orbits when two of the four points coincide. When $(a, b, c, d) = -1$ we say that the four points are in *harmonic division* (or that they form a *harmonic quadrilateral* if $K = \mathbf{C}$); four points in harmonic division are distinct in characteristic $\neq 2$. The cross-ratios $-j$ and $-j^2$ are sometimes called *equianharmonic*.

The permutations that leave invariant the cross-ratio -1 form a subgroup of $\mathcal{S}_4$ of order $8 = 24/3$, generated by $\mathbf{H}$ and the transposition $(1, 2)$.

The subgroup of permutations leaving invariant the cross-ratio $-j$ has $24/2 = 12$ elements; it is the alternating group $\mathbf{A}_4$.

Theorem 27. *Assume* K *has characteristic* $\neq 2$.

(a) $(a, b, c, d) = -1$ *is equivalent to* $2(ab + cd) = (a + b)(c + d)$;

(b) $(a, b, c, \infty) = -1$ *is equivalent to* $2c = a + b$;

(c) $(a, -a, c, d) = -1$ *is equivalent to* $cd = a^2$.

Proof. By theorem 21, $(a, b, c, d) = -1$ is equivalent to

$$(c-a)(d-b) + (c-b)(d-a) = 0,$$

and (a) follows after a short calculation. From (a) we immediately derive (b) and (c), using the conventions on ∞ in the first case. □

The *harmonic conjugate* of c with respect to a and b is the point d such that $(a, b, c, d) = -1$. When the projective line D is embedded in a projective plane, the figure below shows how to find this point using only a ruler.

First join a and b to a point $m \notin D$, and draw a line through c intersecting D_{ma} and D_{mb} at points a' and b' distinct from m. Setting $p = D_{ab'} \cap D_{ba'}$ and intersecting D_{mp} with D gives d. To justify this construction, choose

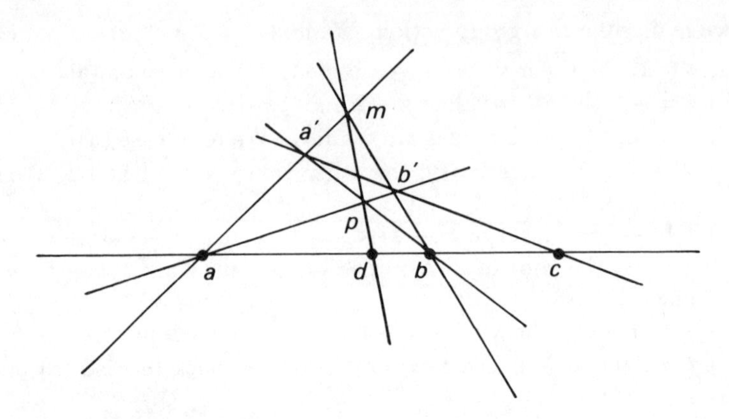

a projective frame whose vertices are a, b, m and such that D_{am} is the line at infinity; if $(-u, 0)$ are the affine coordinates of c and $(0, v)$ those of b', the equation of $D_{cb'}$ is $y = (x + u)u^{-1}v$, and that of D_{bp}, which is parallel to it, is $y = xu^{-1}v$. Since the second coordinate of p is v, the first must be u, implying that the coordinates of d are $(u, 0)$ and that $(a, b, c, d) = (\infty, 0, -u, u) = -1$ by theorem 27(b).

Theorem 28. *Let K be a field of characteristic $\neq 2$ and $f : \hat{K} \to \hat{K}$ a bijection. The following properties are equivalent:*

(a) f *is a* semi-projective transformation, *that is, the composition of a projective transformation with an automorphism s of K (in other words, f is of the form $(as(t) + b)/(cs(t) + d)$ for $t \in \hat{K}$);*

(b) f *preserves harmonic divisions.*

Proof. If s is an automorphism of K, the rationality of formula (41) shows that

$$(s(a), s(b), s(c), s(d)) = s((a, b, c, d)). \tag{45}$$

Since projective transformations preserve cross-ratios (theorem 20) and $s(-1) = -1$, (a) implies (b). Conversely, after composing f with a projective transformation, if necessary, we may assume that f leaves invariant ∞, 0 and 1. By theorem 27(b), f preserves midpoints:

$$f\big(\tfrac{1}{2}(a + b)\big) = \tfrac{1}{2}\big(f(a) + f(b)\big);$$

since $f(0) = 0$, we deduce, for $b = 0$, that $f(a/2) = f(a)/2$ for every a. This implies additivity, again because midpoints are preserved. On the other hand, theorem 27(c) shows that $(a, -a, a^2, 1) = -1$, so f preserves squares: $f(a^2) = f(a)^2$ for every $a \in K$. Finally, the identity $4xy = (x + y)^2 - (x - y)^2$, together with the additivity of f, shows that $f(4xy) = 4f(x)f(y)$, whence $f(xy) = f(x)f(y)$. Thus f is multiplicative, hence an automorphism of K. □

Corollary. *If $K = \mathbf{Q}$, $\mathbf{R}$ or $\mathbf{F}_p$, for p an odd prime, every bijection $f : \hat{K} \to \hat{K}$ perserving harmonic divisions is a projective transformation.*

Proof. It is enough to see that the only automorphism s of K is the identity. From $s(1) = 1$ we get by additivity and induction over n that $s(n \cdot 1) = n \cdot 1$ for every $n \in \mathbf{N}$; hence the assertion for $\mathbf{F}_p$. From $s(m) = m$ for $m \in \mathbf{Z}$ we get, by the multiplicativity of s,

$$s(n/m) = s(n)/s(m) = n/m,$$

showing the assertion for $\mathbf{Q}$. Finally, notice that an element of $\mathbf{R}$ is a square if and only if it is positive; thus s takes positive reals into positive reals, so it must be increasing. Since the restriction of s to $\mathbf{Q}$ and $\mathbf{Q}$ is dense in $\mathbf{R}$ with respect to the order structure, s is the identity on $\mathbf{R}$. □

The ruler construction of the harmonic conjugate, together with theorem 28, shows that a collineation of the projective plane induces semi-projective transformations on the lines of the plane. This also follows from the fundamental theorem of projective geometry (theorem 7, section 1.3).

2.4. Projective Transformations and Involutions on a Projective Line

Let $h : D \to D$ be a projective transformation from a projective line onto itself. As we have seen, h can be expressed in the form

(46) $\qquad t' = (at + b)/(ct + b) \qquad$ or $\qquad ctt' + dt' - at - b = 0,$

t and $t' = h(t)$ are projective coordinate ratios on D and a, b, c, d are scalars with $ad - bc \neq 0$. We say that h is an involution if $h \neq 1$ and $h \circ h = 1$; this means that t and t' are interchangeable in (46), that is, $d = -a$. If h is an involution and $t' = h(t)$, we say that t and t' are *homologous*.

Theorem 29. *Let $h : D \to D$ be a projective transformation of a projective line. The following assertions are equivalent:*

(a) *h is an involution;*

(b) *h is of the form $\mathbf{P}(u)$, where u is a linear map with trace zero;*

(c) *there exists a point $m \in D$ such that $h(m) \neq m$ and $h^2(m) = m$.*

Proof. In the notation of (46), the linear maps u such that $h = \mathbf{P}(u)$ have matrices proportional to $M = \left(\begin{smallmatrix} d & c \\ b & a \end{smallmatrix}\right)$. Since $\operatorname{tr} M = d + a$, (a) and (b) are equivalent. It is clear that (a) implies (c). If (c) is true, we can take m and $h(m)$ as ∞ and 0, so $h(\infty) = 0$ and $h(0) = \infty$, and $a = d = 0$ in (46), which means $tt' = b/c$, a symmetric relation. □

Corollary. *An involution h is uniquely determined by two pairs (p, p') and (q, q') of non-fixed homologous points.*

Proof. The unique projective transformation h such that $h(p) = p'$, $h(p') = p$ and $h(q) = q'$ is an involution by theorem 29(c). □

In the notation of (46) the fixed points of the projective transformation h have as projective coordinate ratios the (finite or infinite) roots of the equation

$$ct^2 + (d - a)t - b = 0. \tag{47}$$

If $h \neq 1$, there exist at most two such fixed points.

Theorem 30. *If a projective transformation $h : D \to D$ has two distinct fixed points p and q, the cross-ratio $(p, q, m, h(m))$ is a constant k not depending on the point $m \in D$. This constant is the ratio between the eigenvalues of the linear map u such that $h = \mathbf{P}(u)$. The projective transformation h is an involution if and only if $k = -1$, that is, if and only if $p, q, m, h(m)$ are in harmonic division.*

Proof. Take p and q as the first two elements of a projective frame. In terms of projective coordinate ratios, h has the form $h(t) = kt$. Thus $(p, q, m, h(m)) = (\infty, 0, t, kt) = k$. The matrix of u can be taken equal to $\left(\begin{smallmatrix} 1 & 0 \\ 0 & k \end{smallmatrix}\right)$, whence the assertion on the eigenvalues. Finally, $h(t) = kt$ is an involution if and only if $k = -1$. □

Reduced forms

If a projective transformation has two fixed points, we have just seen that, in an appropriate projective coordinate system, its expression is

$$t' = kt \qquad \text{(homothety)}. \tag{48}$$

If it has a double fixed point and we place it at infinity, the expression is

$$t' = t + b \qquad \text{(translation)}. \tag{49}$$

When h is an involution ($d = -a$), equation (47) can be written $ct^2 - 2at - b = 0$. Since $a^2 + bc = bc - ad \neq 0$, its roots are distinct in characteristic $\neq 2$, and the reduced form of the involution, given by (48), is $t' = -t$ (reflection). In characteristic 2, the equation has a double root and the involution has, in an appropriate base, the expression $t' = t + b$ (or, equivalently, $t + t' = b$).

The discussion above assumes that the roots of (47) are in K, or that it's OK to replace K by a quadratic extension. Without that hypothesis other reduced forms have to be sought. Start from a non-fixed point p of the projective transformation h, and take p and $h(p)$ as elements of a projective frame (that is, $h(0) = \infty$). Then h has a formula of the form

$t' = (at+b)/t$. If h is an involution, we have $h(\infty) = 0$, whence $a = 0$ and the reduced form

$$t' = b/t \qquad \text{or} \qquad tt' = b. \tag{50}$$

If h is not an involution, $h(\infty)$ is distinct form 0 and ∞, and we take it to be 1. Thus $a = 1$, and we get the reduced form

$$t' = (t+b)/t. \tag{51}$$

Formula (50) shows that the fixed points of an involution coincide in characteristic 2, and are distinct otherwise (since $b \neq 0$). In equation (51), the fixed points are given by the equation $t^2 - t - b = 0$; they are always distinct in characteristic 2, which means that in characteristic 2 involutions are exactly those projective transformations having a double fixed point.

Theorem 31. *Every projective transformation $h : D \to D$ is the product of at most two involutions.*

Proof. We use (51): setting $t'' = -b/t$, we get $t' + t'' = 1$. □

Involutions and divisors of degree two

We defined in section 1.8 the divisors of a projective line D as the formal linear combinations of points of D. Divisors of degree n can then be seen as the orbits of the symmetric group $\mathcal{S}_n$ acting on D^n by permutation of the factors.

We recall that the degree-two divisors of D form a two-dimensional projective space: the sum of the points with homogeneous coordinates (x, y) and (x', y') has homogeneous coordinates $(xx', -(xy' + yx'), yy')$. If we parametrize the two points of D by $t = y/x$ and $t' = y'/x'$, which we assume finite, we can take $(1, -(t+t'), tt')$ as the homogeneous coordinates of they sum. Since an involution j on D is expressed by a relation of the form $att' + b(t+t') + c = 0$, we can see it as a pencil of degree-two divisors, the divisors being $\{m, j(m)\}$ for all $m \in D$.

To be exact we should assume K algebraically closed, since, for given tt' and $t + t'$ in K, the roots t and t' are in a quadratic extension of K.

For $a \in D$, the divisors $a + m$, for m running over D, form a pencil, which we can call *degenerate*, and which does not correspond to a true involution. □

A pencil of degree-two divisors corresponds also to a linear pencil of quadratic equations. Whence the following result:

Theorem 31 (Desargues-Sturm). *A pencil $\mathcal{F}$ of conics of a projective plane determines an involution on every line D that does not intersect the base of $\mathcal{F}$.*

Proof. The intersection of a conic C with D is described by a quadratic equation whose coefficients are linear functions of the coefficients of the equation of C. The assumption on the base is designed to avoid degenerate pencils. □

Here's an alternative proof: Each $m \in D$ is contained in exactly one conic C of $\mathcal{F}$, which intersects D again at a point $\jmath(m)$ (section 1.7). The projective coordinate ratio of $\jmath(m)$ is a rational function of that of m. Furthermore, $\jmath(\jmath(m)) = m$. The corollary to theorem 24 (section 1) completes the proof.

Projective transformations and involutions on a pencil of lines

A pencil of lines is the set of lines that pass through a point of the projective plane. The projective coordinate ratio of a line D in such a pencil is often called its *slope*.

In particular, let us consider the projective closure of a Euclidean plane and the pencil $\mathcal{F}$ of lines that contain a fixed point O of the plane (cf. secion 1.6). In an orthonormal frame, the slope of a line D of $\mathcal{F}$ is $\tan V$, where V is the angle between the x-direction and D. Let I and J be the isotropic lines of O; we propose to compute the cross-ratio (I, J, D, D'). Formula (41) leads us to calculating $(\tan V - i)/(\tan V + i)$ and its analogue for D'. Still denoting by V the angle between the positive x-direction and one of the half-lines determined by D, we get the value

$$(\sin V - i \cos V)/(\sin V + i \cos V) = e^{-2iV}$$

for this ratio, and the desired cross-ratio is equal to $e^{2i(V'-V)}$. But $V' - V$ is the angle W between one of the two half-lines determined by D and one of the half-lines determined by D', and $2W$ does not depend on the choice of half-lines. Thus we get *Laguerre's formula:*

$$(I, J, D, D') = e^{2iW} = \cos 2W + i \sin 2W. \tag{52}$$

From this and theorem 30 we get:

Theorem 32. *In the Euclidean plane, a projective transformation h of a pencil of lines $\mathcal{F}$ that leaves fixed the isotropic lines is induced by a rotation. The angle of this rotation is a right angle if and only if h is an involution, called the orthogonality involution.* □

Thus two lines are orthogonal if and only if they are conjugate with respect to the isotropic lines; and in fact $(i, -i, m, m') = -1$ is equivalent to $mm' = -1$ (theorem 27(c)).

Application

Let a, b, c be a triangle on the Euclidean plane and d the intersection of two of its altitudes. Let G be the pencil of conics containing a, b, c, d

(section 1.7). The degenerate conics $D_{ad} + D_{bc}$ and $D_{bd} + D_{ac}$ are in G. The involution determined by G on the line at infinity (theorem 31) has two pairs which represent orthogonal lines; thus the same is true of all pairs (corollary to theorem 29, and theorem 32). This shows the well-known fact that the three altitudes of a triangle are concurrent, because the lines D_{cd} and D_{ab} that make up the third degenerate conic of G are also orthogonal. Furthermore, every conic contaning a, b, c, d has orthogonal asymptotes; if the asymptotes are real, we say that the conics are *equilateral hyperbolas*.

A construction

Given three concurrent lines A, B, C on an affine plane over a field of characteristic $\neq 2$, the line D conjugate to C with respect to A and B can be constructed in the following way: draw a parallel C' to C; it intersects A and B at a and b. Then D is the line joining the common point of A, B, C with the midpoint c of a and b. Indeed, c is the harmonic conjugate of the point at infinity of C' with respect to a and b (theorem 27(b)).

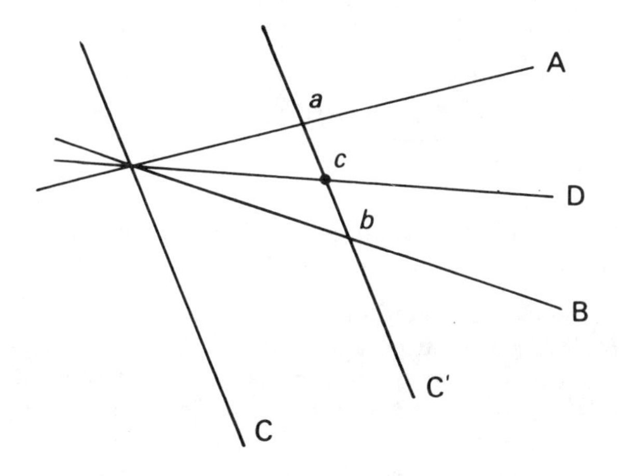

If the plane is Euclidean and C and D are orthogonal, this shows that A and B are symmetric with respect to C (or D, which boils down to the same); in other words, C and D are the bisectors of A and B.

But the identification of involutions with pencils of degree-two divisors, discussed above, shows that two (distinct) involutions on the same line share exactly one pair. Thus an involution on a pencil $\mathcal{F}$ of lines of the Euclidean plane has one orthogonal pair. The fixed lines of such an involution are symmetric with respect to either of the two orthogonal lines.

2.5. The Projective Structure of a Conic

Let C be an irreducible conic in a projective plane, and $a \in C$ a point. To each line D of the pencil F_a of base a, we associate the second point where D intersects C; we denote this point by $j_a(D)$. When D is the tangent to C at a, let $j_a(D)$ be the point a. Thus j_a is a bijection from F_a to C. It allows us to transport to C the projective line structure of F_a: the "slope" t of a line D through a is taken as the projective coordinate ratio of the point $j_a(D)$ of C. There remains to see that this structure on C does not depend on the point a. This follows from the next theorem:

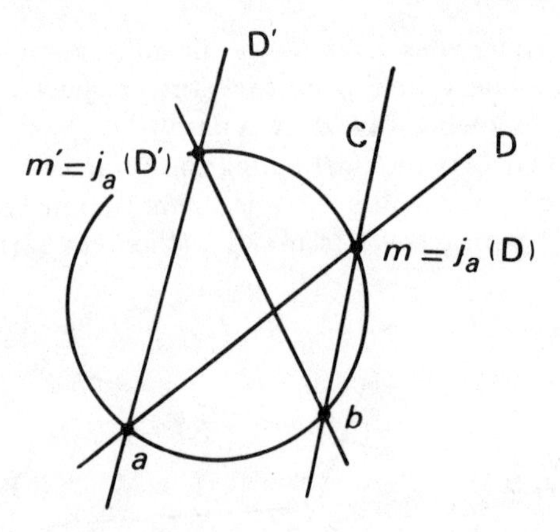

Theorem 33. *With the notations above, if b is another point on C, the composition $j_b^{-1} \circ j_a$ is a projective transformation from F_a to F_b.*

Proof. For $D \in F_a$, the coordinates of $j_a(D)$ are rational functions of the slope t of D because, in parametrizing D, $j_a(D)$ corresponds to the second root of a quadratic equation whose first root—the parameter of a—is known. Thus the slope t' of the line $j_b^{-1}(j_a(D))$, which joins b and $j_a(D)$, is also a rational function of t. Conversely, t is a rational function of t'. The result now follows from theorem 24 (section 1). □

For those who are wary of such "pure-thought" proofs, here's a more explicit one. Take a projective frame for the plane in which the homogeneous coordinates of a and b are $(1,0,0)$ and $(0,1,0)$. The general equation of the conics containing a and b is then

$$pz^2 + qxy + ryz + sxz = 0. \tag{53}$$

Let $y = tz$ and $x = t'z$ be the equations of the lines D and D', going through a and b, respectively. The intersection of D and D' is on C if and only if $p + qtt' + rt + st' = 0$; this defines t' as a degree-one rational function of t, and vice versa.

Theorem 33 says that the bijection j_a gives a parametric representation for the conic C: if we denote by t the slope of a variable line going through a, the affine coordinates of the point $j_a(D)$ are rational functions of t. For example, in the notation of the second proof above, $j_a(D)$ has affine coordinates

$$x = -(rt + p)/(qt + s) \qquad \text{and} \qquad y = t;$$

its homogeneous coordinates can be taken to be

$$(-(rt + p), t(qt + s), qt + s). \tag{54}$$

More generally, a plane curve that admits an affine parametric representation of the form $x = r(t)$, $y = s(t)$, where $r(t)$ and $s(t)$ are rational functions of a parameter t (which runs through $\hat{K}$) is called *unicursal* or *rational*. After clearing denominators, we get for such a curve a "homogeneous parametric representation", where the homogeneous coordinates of a point of the curve are polynomial functions of t:

$$(x, y, z) = \big(P(t), Q(t), R(t)\big);$$

we assume that the three polynomials P, Q and R do not share a common factor.

The homogeneous parametric representation (54) shows that a conic C is a unicursal curve, for which $\max(d^0P, d^0Q, d^0R) = 2$. This remains true irrespective of the choice of a frame (since P, Q and R are replaced by a linear combination of themselves), and also of the parameter (if t changes by a projective transformation, its "homogeneous coordinates" (u, v) change by a linear map).

The invariance of the maximum degree of P, Q and R is true for every unicursal curve. We will see in the next section that this integer can, with one caveat, be interpreted as the degree of the curve C.

Theorem 33 admits the following converse:

Theorem 34. *In a projective plane, let F_a and F_b be pencils of lines with base points a and b, and $h : F_a \to F_b$ a projective transformation. As D runs through F_a, the point $D \cap h(D)$ describes a conic C, which contains a and b. If $h(D_{ab}) = D_{ba}$, the conic C degenerates into D_{ab} and another line; otherwise C is non-degenerate.*

Proof. We use again the notation (53) of the second proof of theorem 33. The conic $pz^2 + qxy + ryz + sxz = 0$, which goes through a and b, is obtained by the projective transformation $t' = -(rt + p)/(qt + s)$. Every projective transformation from F_a onto F_b is of this form, with $pq - rs \neq 0$. The conic cannot split into a line through a and one through b, or we'd have $pq - rs = 0$; it can only split into D_{ab} (whose equation is $z = 0$) and another line. This implies that $q = 0$, and that ∞ is fixed under the projective transformation. This means that D_{ab} is taken to itself by the projective transformation. □

Example. Take for a and b two points in the Euclidean plane and for h the composition of the translation from a to b with a rotation of angle W around b. The conic C is then the set of points from which one sees the segment ab under an angle (of lines) W. Since h takes the isotropic line of a into the parallel isotropic line at b (see section 4), the conic C goes through the cyclic points, and is a circle containing a and b.

Theorem 35 (Frégier). *Let C be an irreducible conic and $f \notin C$ a point in the plane; the map j which, to every point $m \in C$, associates the point $D_{fm} \cap C$ is an involution on C. Conversely, every involution k on C is of this form.*

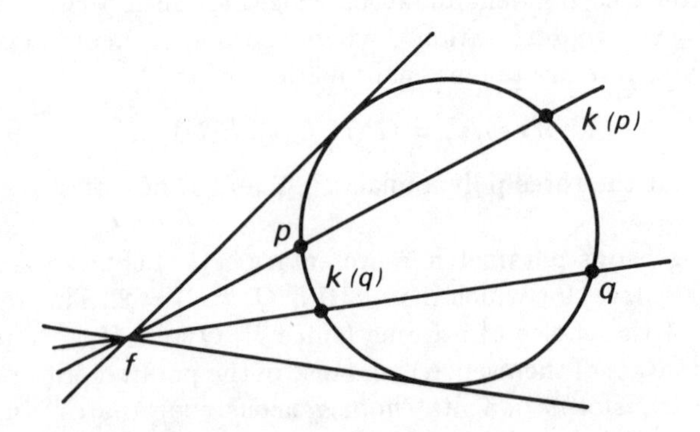

Proof. In fact, j is a rational map from C into itself, and clearly $j^2 = 1$. Thus j is a projective transformation, and an involution, by the corollary to theorem 24. Conversely, if k is an involution on C, take two non-homologous points p and q on C and set $f = D_{pk(p)} \cap D_{qk(q)}$; the involution defined by f has two pairs of homologous points in common with k, hence it coincides with k by the corollary to theorem 29. □

A calculus proof would also be quite simple. Let $(x, y, z) = \big(P(t), Q(t), R(t)\big)$ be a homogeneous parametric representation for C. The polynomials P, Q and R are linearly independent (otherwise C would be a line) and thus form a basis for $K + Kt + Kt^2$. By changing the projective coordinate system, we can assume that $x = t^2$, $y = t$ and $z = 1$. Then the fact that the points of C with parameters t and t' are collinear with the point $f = (a, b, c)$ is expressed by the equality

$$\begin{vmatrix} a & t^2 & t'^2 \\ b & t & t \\ c & 1 & 1 \end{vmatrix} = 0,$$

which, after divison by $t - t'$, becomes $a - b(t + t') + ctt' = 0$. This is indeed an involution on C, and, for some choice of (a, b, c), every involution on C is obtained in this way.

The point f around which the lines $D_{mj(m)}$ revolve is called the *Frégier point* of the involution j.

Remark. The involution $t' = (bt - a)/(ct - b)$ is "degenerate" if $b^2 - ac = 0$; this means that $(a, b, c) \in C$.

Examples

(1) Consider, in characteristic 2, the involution $t + t' = 0$ on a conic C. This involution is the identity, so it is determined on C by the tangents to C. Thus all tangents to C go through the Frégier point of this involution (cf. section 1.7).

(2) Let a be a point on a conic C in the Euclidean plane. The orthogonality involution (theorem 32, section 4) on the pencil of lines going through a induces an involution on C. More exactly, if a right angle rotates around a and we call m and m' the points where its sides intersect C, the line mm' goes through a fixed point f. By taking the tangent to C at a as one of the sides of this right angle, we conclude that the Frégier point f of this involution lies on the normal to C at a. This was Frégier's original statement (Ann. de Math., 1816).

If C is a circle, f is clearly its center.

If C is an equilateral hyperbola (one with orthogonal asymptotes), the line at infinity is a particular case of the line mm', so the Frégier point f is at infinity on the normal to C at a. Thus the involution is determined on C by parallels to this normal.

Conversely, consider the involution j determined on an equilateral hyperbola C by the lines parallel to a given line D. Let a and b be the points of C where the tangent is orthogonal to D. Then the orthogonality involution on the pencil of lines with base a induces j on C, and similarly for the pencil with base b. In other words, the circles whose diameters are the chords mm' of C parallel to D all contain the points a and b, hence they form a pencil of circles.

The Frégier point $f(a)$ of the involution determined on a conic C by the orthogonality involution on the pencil of lines going through a point a of C can't be just any point, because it is a (rational) function of one parameter (the parameter of a on C), not two. We have seen that if C is a circle $f(a)$ is the center of C and if C is an equilateral hyperbola $f(a)$ describes the line at infinity. Otherwise it describes a curve that is, in general, of the fourth degree.

(3) The fixed points of an involution on a conic C are the contact points of the tangents to C that contain the Frégier point.

(4) The existence of a unique pair common to two involutions is immediate if one considers them as involutions on the same conic: just take the line joining their Frégier points.

Theorem 36 (Pascal). *Given6 points* $1, 2, 3, 4, 5, 6$ *on an irreducible conic* C, *the points* $i = D_{12} \cap D_{45}$, $j = D_{23} \cap D_{56}$ *and* $k = D_{34} \cap D_{61}$ *are collinear.*

Lemma. *Let* C *be an irreducible conic,* D *a line,* u *and* v *its intersections with* C, *and* a, b, a', b' *points on* C. *There exists a projective transformation*

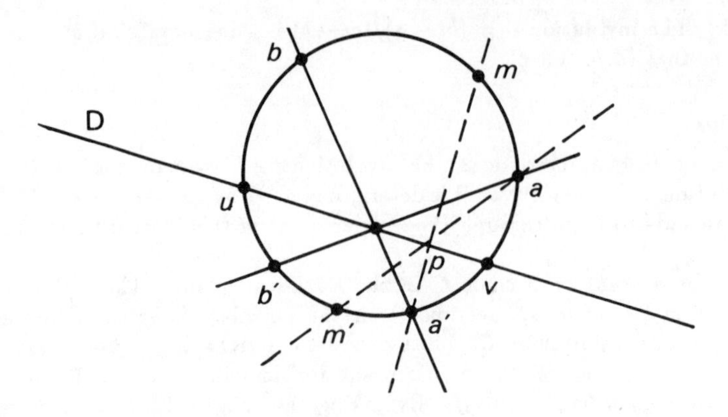

h on C fixing u and v and taking a and b to a' and b', respectively, if and only if the intersection $D_{ab'} \cap D_{ba'}$ belongs to D.

Proof. For every $m \in C$, let p be the point where $D_{ma'}$ meets D and m' the point where D_{ap} intersects C again, and set $m' = h(m)$. The map h is rational and has an inverse that is also rational; thus h is a projective transformation. We have $h(a) = a'$ and m is fixed by h if and only if m belongs to both C and D. If $D_{ab'}$ and $D_{ba'}$ intersect on D, it is clear that $h(b) = b'$. The converse is true because a projective transformation is determined by where it takes three points (here u, v and a). □

If D is tangent to C at u, u is a double fixed point of h. The lemma is still true because in this case h is determined by its value somewhere else (cf. the reduced form $t' = t + b$, equation (49) in section 4).

When the lines ab' and ba' intersect on D, there is also a projective transformation k with fixed points u and v and such that $k(a) = b$ and $k(a') = b'$.

Proof of theorem 36. Let u and v be the points where C intersects $D = D_{ij}$. By the lemma, there exists a projective transformation h on C fixing u and v and such that $h(2) = 5$; then $h(1) = 4$ and $h(3) = 6$. By using the lemma the other way around, one sees that D_{16} and D_{34} intersect on D. Thus i, j and k are collinear. □

The order of the six points matters only up to circular permutations. If the points occur in the order $1, 2, 3, 4, 5, 6$ on an ellipse, parabola or branch of real hyperbola, they form a convex hexagon inscribed in the conic and theorem 36 says that the three intersection points of pairs of opposite sides are collinear.

When C degenerates into two lines, with $1, 3, 5$ on one and $2, 4, 6$ on the other, we recover Pappus's theorem (theorem 6 in section 1.3), but the proof given here is not valid in this case.

Corollary.

(a) *If a, b, c, d are points on an irreducible conic C, the tangents to C at a and c intersect on the line that joints the points $D_{ab} \cap D_{cd}$ and $D_{bc} \cap D_{da}$.*

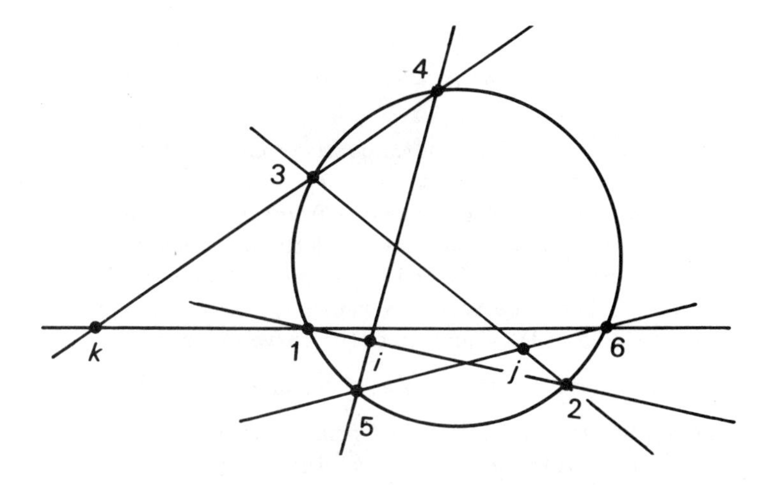

(b) *If a triangle a, b, c is inscribed in an irreducible conic C, the intersection of the tangents at a, b, c with the opposite sides are collinear.*

Proof. For (a), consider the sextuple a, a, b, c, c, d. For (b), consider a, a, b, b, c, c. □

If C is the circle circumscribed to a, b, c we obtain *Simpson's line*.

2.6. Unicursal Curves

We recall that a unicursal curve is one that admits an (affine) parametric representation of the form $x = r(t)$, $y = s(t)$, where $r(t)$ and $s(t)$ are rational functions of a parameter t. This notion can be immediately generalized to curves in K^n or $\mathbf{P}_n(K)$.

Examples

(1) *Every plane algebraic curve of degree d having a multiple point of multiplicity $d-1$ is unicursal.*

Place the multiple point at the origin, so that the affine equation of the curve is of the form $F_d(x,y) + F_{d-1}(x,y) = 0$, where F_j is a homogeneous polynomial of degree j. Set $y = tx$, that is, consider the intersection of the curves with lines through the origin. The equation gives

$$x^d F_d(1,t) + x^{d-1} F_{d-1}(1,t) = 0.$$

The solution $x^{d-1} = 0$ corresponds to the origin. There remains the point $x = -F_{d-1}(1,t)/F_d(1,t)$, whose y-coordinate is

$$tx = -tF_{d-1}(1,t)/F_d(1,t).$$

This is the desired parametric representation.

For $d = 2$ we recover the parametric representations of conic studied in section 5.

(2) *Every quartic C with three double points is unicursal.*

Recall that a quartic is a curve of degree 4. Take the three double points as the vertices of a projective frame for the plane. The fact that $(0, 0, 1)$ is a double point says that the homogeneous equation of C has no terms in z^4, in z^3x or in z^3y, and similarly for the other two variables. In short, the variables x, y, z appear in the equation with exponents ≤ 2, and the equation of the curve reduces to

$$ax^2y^2 + by^2z^2 + cz^2x^2 + a'xyz^2 + b'yzx^2 + c'zxy^2 = 0.$$

Now apply the *quadratic transformation* $x = 1/x'$, $y = 1/y'$, $z = 1/z'$. By multiplying the resulting equation by $(x'y'z')^2$, we obtain

$$az'^2 + bx'^2 + cy'^2 + a'x'y' + b'y'z' + c'z'x' = 0.$$

This is the equation of a conic, which we parametrize as in section 5. From the homogeneous parametric representation $(x', y', z') = \big(P(t), Q(t), r(t)\big)$ of this conic we deduce one for the quartic, namely

$$(x, y, z) = \big(Q(t)R(t), R(t)P(t), P(t)Q(t)\big).$$

The equality sign for homogeneous parametric representations stands, of course, for proportionality.

A parametric representation $x = r(t)$, $y = s(t)$ of a unicursal curve is said *proper* if the parameter t is a rational function of x and y. Then, except for a finite number of exceptional points where $t = p(x, y)/q(x, y)$ has the form $0/0$, there is a bijection between the curve and the set $\hat{K}$ of values of the parameter.

Examples

(1) The parametric representation above of a degree-d curve having a point of multiplicity $d - 1$ is proper, since $t = y/x$. The values of t such that $F_{d-1}(1, t) = 0$ all give the origin; in general there are several such values. The values of t such that $F_d(1, t) = 0$ give the points at infinity of the curve; don't forget that t can also take the value ∞. The reader can digest these ideas by studying the curves $x^3 - 2x^2y - xy^2 + 2y^3 + xy = 0$ and $x^3 - y^2 = 0$.

The parametric representation above of the quartic with three double points is also proper, since the parametric representation of a conic is proper.

(2) Let C be the curve whose parametric representation is $x = t^2 + t^{-2}$, $y = t^3 + t^{-3}$. This representation is not proper because two values

t and $t' = t^{-1}$ of the parameter yield the same point (x, y). But, if we set $u = t + t^{-1}$, we have $x = u^2 - 2$ and $y = u^3 - 3u$. Since $u^3 = u \cdot u^2 = u(x+2)$, we deduce that $y = u(x+2) - 3u = u(x-1)$; hence $u = y/(x-1)$ and the new parametric representation is proper.

Notice that $u = 0/0$ for $x = 1$ and $y = 0$; this point is obtained for $u = \sqrt{3}$ and $u = -\sqrt{3}$, and is thus a double point of the curve. Notice also that the old parameter t is not a rational function of the new parameter u; it is quadratic over $K(u)$.

What we saw in example (2) is general: every representation can be made proper. More precisely, if t is considered as an indeterminate over K, $x = r(t)$ and $y = s(t)$ are elements of the field of rational fractions $K(t)$. Together they generate a subfield $F = K(x, y)$ of $K(t)$, so the point is to show that F is generated (over K) by a single element u, that is, $F = K(u)$. If this is true, x and y are rational functions of the new parameter u, and u is a rational function of x and y. Now we have the following theorem:

Theorem 37 (Lüroth). *Let $E = K(t)$ be a field of rational fractions in one variable over K, and $F \neq K$ a field between K and E. Then F is of the form $F = K(u)$, with u transcendent over K. In particular, every unicursal curve has a proper representation.*

Proof. Since F contains a non-constant rational fraction, E is an algebraic extension of finite degree of F (theorem 23). Let n be the degree of this extension, and $f(X) = X^n + k_1X^{n-1} + \cdots + k_{n-1}X + k_n$, for $k_j \in F$, the minimal polynomial of t over F. At least one of the k_j is not in K (otherwise t would be algebraic over K); denote it by u. We will see that $F = K(u)$.

By writing $u = g(t)/h(t)$ as the quotient of relatively prime polynomials, and setting $m = \max(d^0g, d^0h)$, we see that E is of degree m over $K(u)$ (theorem 23). Since $K(u)$ is contained in F, we have $m \geq n$; in fact, m is a multiple of n. There remains to show that $m \leq n$.

The indeterminate t is a root of the polynomial $g(X) - uh(X)$, which has coefficients in F. Thus this polynomial is a multiple of the minimal polynomial $f(X)$:

(E) $$g(X) - uh(X) = q(X)f(X), \qquad \text{with } q(X) \in F[X].$$

Each k_j can be written in the form $c_j(t)/c_0(t)$, where $c_0, c_1, \ldots, c_n$ are polynomials without common factors. We now set

(F) $$P(X,t) = c_0(t)f(X) = c_0(t)X^n + c_1(t)X^{n-1} + \cdots + c_n(t);$$

by assumption P is a primitive polynomial in X over $K[t]$.

Since $u = k_i$ for some i, we have $u = g(t)/h(t) = c_i(t)/c_0(t)$, and $d^0g \leq d^0c_i$, $d^0h \leq d^0c_0$ because $(g, h) = 1$. Thus,

(G) $$m = \max(d^0g, d^0h) \leq \text{degree of } P(X,t) \text{ with respect to } t.$$

Now, in (E), replace u by $g(t)/h(t)$ and the coefficients of $q(X)$ by their expressions as rational functions of t. It follows that, in $K(t)[X]$, the element $P(X,t)$ divides $g(X)h(t) - h(X)g(t)$. But these polynomials are all in $K[t,X]$, and we've seen that $P(X,t)$ is primitive in X. Thus there exists a polynomial $Q(X,t) \in K[X,t]$ such that

$$g(X)h(t) - h(X)g(t) = Q(X,t)P(X,t). \tag{H}$$

The degree of the left-hand side with respect to X is at most m, and the degree of $P(X,t)$ is $n \leq m$. Thus both sides have degree m, and $n = m$. □

Furthermore $Q(X,t)$ has degree 0 in X, that is, it depends only on t. Since the left-hand side of (H) changes only by a sign if we switch the indeterminates X and t, it follows that $Q(X,t)$ is a constant, and that $P(X,t)$ is also of degree $m = n$ in t.

Remark. The generalization of Lüroth's theorem to the field of rational fractions in several variables is a problem of longstanding interest to algebraic geometers. Around 1900, using the powerful tools provided by the theory of algebraic surfaces, G. Castelnuovo proved that, if $K = \mathbf{C}$, every subfield F of transcendence degree 2 of $K(X,Y)$ (that is, over which $K(X,Y)$ is algebraic) is isomorphic to $K(X,Y)$ (we say that F is a *pure transcendent extension* of K). The conclusion is not true when K is not algebraically closed: B. Segre found a counterexample for $K = \mathbf{R}$. It is still true over an algebraically closed field K of characteristic $p \neq 0$, as long as $K(X,Y)$ is separable over F (O. Zariski, around 1955).

By contrast with Lüroth's theorem, there is no known purely field-theoretic proof for the theorems of Castelnuovo and Zariski. Their proofs all resort to algebraic geometry. A (d-dimensional) algebraic variety V is called *unirational* if the (say, affine) coordinates of points of V can be expressed as rational functions of d independent parameters t_j; it is called *rational* if in addition one can choose the parameters as rational functions of the coordinates. As we have seen, Lüroth's theorem implies that every unirational curve is rational. The proofs of Castelnuovo and Zariski consist in showing that certain invariants of a unirational surface vanish, and that this fact implies the rationality of the surface (which is then *birationally equivalent*, that is, isomorphic in a certain sense, to the projective plane).

It was not until the seventies that convincing examples were found of unirational three-dimensional varieties which are not rational. One of them is Fano's cubic hypersurface ($F(x,y,z,t) = 0$, for a fairly general cubic polynomial F), whose unirationality had been known since the beginning of the century.

Theorem 38. *A(plane) unicursal curve C is algebraic and irreducible.*

Proof. We can argue in the affine setting. Let $x = r(t)$, $y = s(t)$ be a proper representation for C, so t is a rational function $t = G(x,y)$ of x and y. In the field $K(t)$ of rational fractions, $x = r(t)$ is non-constant (else C is a straight line), so x is transcendent over K, and y is algebraic over $K(x)$

(theorem 23). Let $F(x, Y)$ be the minimal polynomial of y over $K(x)$; we can assume that the coefficients of each Y^j are elements of $K[x]$, that is, polynomials in x, and that they have no common factors. Being irreducible over $K(X)$ and primitive over $K[X]$, the polynomial $F(X, Y)$ is irreducible. Since $F(r(t), s(t))$ is an identity in t, every point $(x(a), y(a))$ of C, for $a \in \hat{K}$, satisfies $F(x(a), y(a)) = 0$. Conversely, if (b, c) is a zero of $F(X, Y)$ in K^2, we set $a = G(b, c)$; then the identity $t = G(x, y) = G(r(t), s(t))$ shows that the K-homomorphism h of $K[x, y]$ into K defined by $h(x) = b$, $h(y) = c$ can be extended to t by setting $h(t) = a$. Thus $b = h(r(t)) = r(a)$ and $c = s(a)$, so that the point (b, c) is on C. If $G(b, c)$ is infinite, we take $t' = 1/t$. □

More precisely, by taking a valuation ring of $K(t)$ containing x and y, we see that the homomorphism h can be extended to t or to $1/t$.

A homogeneous parametric representation $(P(t), Q(t), R(t))$ of a unicursal curve C is called *proper* if the affine parametric representation $x = P(t)/R(t)$, $y = Q(t)/R(t)$ is proper. Up to a finite number of exceptions, such a parametric representation gives a bijection between $\hat{K}$ and the projective curve C. The intersections of a line D of equation $ax + by + cz = 0$ with C correspond to the (finite or infinite) roots of the equation $aP(t) + bQ(t) + cR(t) = 0$. We will take as true that the multiplicities of these roots are equal to the intersection multiplicities of C and D defined in section 1.3 (where C is defined by its equation and D by a parametric representation), it being understood that, if an (exceptional) point P of C occurs for several values of the parameter t, one must sum the corresponding multiplicities in order to obtain the intersection multiplicity of C and D at P. This being so, the number $\max(d^0P, d^0Q, d^0R)$ is the degree of the plane curve C (as long as P, Q and R have no common factors, of course).

We have seen that this number is two for a conic and four for a quartic with three double points.

Remark. The examples given at the beginning of this section might lead one to suppose that a curve is unicursal if and only if it has enough double points. Such is indeed the case: one can show that an irreducible curve C of degree d is unicursal if and only if it has $\frac{1}{2}(d-1)(d-2)$ double points, as long as more complicated multiple points are counted the appropriate number of times. For instance, a triple point counts as three double points; as we saw at the beginning of the section, a quartic with a triple point is unicursal, as is a quartic with three double points. In the general case, deciding upon the "appropriate number of times" involves some fairly sophisticated algebra (the study of integral closures of local rings) and general results on algebraic curves (the notion of genus and the theorem of Riemann–Roch).

A degree-d algebraic curve having more than $\frac{1}{2}(d-1)(d-2)$ (appropriately counted) double points must split into lower-degree curves. For example:

- a conic with a double point is made up of two lines (theorem 15 in section 1.7);
- a cubic with two double points a and b splits into a line and a conic, for the line D joining a and b has four points in common with the cubic, and is thus contained in it;
- a quartic with four double points splits into two conics: for we can draw a conic through these four points that also intersects the quartic elsewhere, for a total of $4 \times 2 + 1 > 8$ intersections, so that the conic is contained in the quartic.

It is not hard to show that a degree-d curve C having $\frac{1}{2}(d-1)(d-2)$ *distinct* double points is unicursal: we can find a pencil of curves whose base consists of the double points (with perhaps some other points of C), and whose elements intersect C at a single variable point. The converse is harder. We will just present a special proof that works for cubics:

Theorem 39. *An irreducible plane cubic C is unicursal if and only if it has a double point.*

Proof. If C has a double point, we take the pencil of lines going through it to obtain a (proper) representation (see example 1 above). Conversely, let $(P(t), Q(t), R(t))$ be a homogeneous parametric representation of C, which we can assume proper (theorem 37). Then $\max(d^0 P, d^0 Q, d^0 R) = 3$, and $V = KP + KQ + KR$ is a subspace of $K + Kt + Kt^2 + Kt^3$ of dimension three, otherwise C is a line. By a change of coordinates, we can replace $(P(t), Q(t), R(t))$ by any basis of V.

The intersection of V with the subspace $K + Kt$ has dimension 2 or 1.

(a) If it has dimension 2, V contains 1 and t and has a basis of the form $(1, t, at^3 + bt^2)$, whence the parametric representation $x = 1$, $y = t$, $z = at^3 + bt^2$; thus $z = ay^3 + by^2$, or, after homogenizing, $zx^2 = ay^3 + bxy^2$. The point $(0,0,1)$ is double.

(b) If the intersection is just K, we can take $x = 1$, y of degree two and z of degree three. Applying a translation along the y-axis (i.e., replacing y by $y + kx$) and an affine transformation to t, we can make $y = t^2$. We can also make $z = t^3 + at$ by multiplying z by a scalar and adding linear combinations of x and y. This gives $z = t(y + a)$, $z^2 = t^2(y+a)^2 = y(y+a)^2$, and finally $z^2 x = y(y + ax)^2$ after homogenizing. The point $z = 0$, $y + ax = 0$, that is, $(1, -a, 0)$, is double.

(c) If the intersection has dimension one but is distinct from K, it is generated by a linear polynomial which, by an affine transformation, can be assumed to be t. Thus $x = t$, y has degree two and z degree three. Multiplying y by a scalar and adding a multiple of x, we can assume $y = t^2 + a$.

If $a \neq 0$, a linear combination of z with x and y allows us to reduce z to the form $t^3 + ct^2$. We pass to the affine coordinates $u = y/x = (t^2+a)/t$ and $v = z/x = t^2 + ct = t(t+c)$. Then we have $ut = v - ct + a$, whence $t = (v+a)/(u+c)$. Set $X = u+c$, $Y = v+a$, whence $t = Y/X$ and $t+c = (Y+cX)/X$; thus $Y - a = v = t(t+c) = Y(Y+cX)/X^2$, whence the affine equation $X^2(Y-a) - Y(Y+cX) = 0$, which shows that the origin is a double point.

If $a = 0$, we reduce as above to $x = t$, $y = t^2$, $z = t^3 + b$. Whence the affine representation $u = y/x = t$, $v = z/x = (t^3+b)/t$ and the equation $uv = u^3 + b$. After homogeneization, $uvw = u^3 + bw^3$, and we see the double point $(0,1,0)$.

□

A bit of the geometry of a unicursal cubic

Let C be a unicursal cubic. If we put the double point at the origin, the equation of C is

$$F(x,y) + G(x,y) = 0,$$

where F and G are homogenous of degree three and two, respectively. Assume now that K is algebraically closed. There are two possibilites for G:

(a) G is the product of two distinct linear forms; we say the double point is a *node*. By a change of coordinates, we can write $G(x,y) = -xy$, whence $F(x,y) - xy = 0$. We parametrize by setting $y = tx$, to get

$$(t, t^2, F(1,t)), \qquad \text{or} \qquad x = t/F(1,t),\ y = t^2/F(1,t). \tag{55}$$

Notice that $F(1,t)$ is indeed of degree three, otherwise x factors out of the equation of C. The intersections of C with the line $ax+by+cz = 0$ correspond to the (finite or infinite, simple or multiple) roots of

$$at + bt^2 + cF(1,t) = 0.$$

The term in t^3 and the constant term of this equation come from $F(1,t)$, where they are non-zero (otherwise x or y would factor out); thus the quotient of their coefficients does not depend on the line $ax + by + cz = 0$. Since this quotient is, up to sign, the product of the roots of the equation, the parameters t, t', t'' of the intersections of C with an arbitrary line satisfy $tt't'' =$ constant, and in fact we can take the constant to be 1 by multiplying t by an appropriate scalar. Conversely, if the parameters t, t', t'' of three points M, M', M'' of C are such that $tt't'' = 1$, the line joining M to M' intersects C at a point whose parameter t''_1 satisfies $tt't''_1 = 1$; thus $t''_1 = t''$ and the points M, M', M'' are collinear.

The double point at the origin corresponds to the values 0 and ∞ of t. The (trivial) fact that it is collinear with any point of C leads us to write $0 \cdot \infty \cdot t = 1$, which is not unreasonable.

(b) If the linear factors of $G(x,y)$ are identical, the double point is a *cusp*. We can assume $G(x,y) = -x^2$, whence the equation $F(x,y) - x^2 = 0$

for C. By setting $y = tx$, we get the parametric representation

$$(56)\qquad (1, t, F(1,t)), \qquad \text{or} \qquad x = 1/F(1,t),\ y = t/F(1,t).$$

The intersections of C with $ax + by + cz = 0$ are given by $a + bt + cF(1,t) = 0$. Here the terms in t^2 and t^3 come only from $F(1,t)$, so the ratio of their coefficients is fixed; this ratio is, up to sign, the sum of the roots of the equation. Thus the parameters t, t', t'' of the points where C intersects an arbitrary line satisfy $t + t' + t'' =$ constant. The converse can be shown as in case (a).

Applying a translation to t we can choose the constant to be zero, except in characteristic 3.

The cusp corresponds to the value ∞ of the parameter. The fact that it is collinear with any point of C leads us to write $\infty + \infty + t = k$, which is not unreasonable (there is only one point at infinity; even on $\mathbf{R}$, it has no sign).

Theorem 40. *For an appropriately chosen parametric representation of a unicursal cubic C the points of parameters t, t', t'' of C are collinear if and only if:*

(a) $tt't'' = 1$ *if C has a node;*

(b) $t + t' + t'' = k$ *if C has a cusp.* □

Here are some applications of this theorem.

(1) The *tangential* of a (simple) point M of C is the point M' where the tangent to C at M intersects C again. If t is the parameter of M, the parameter u of its tangential is then $u = 1/t^2$ or $k - 2t$, according to the cases above. A short calculation shows that the tangentials of three collinear points are collinear. In characteristic two, $k - 2t = k$, so all the tangents to a cuspidal cubic go through the point of parameter k.

(2) Given a point M of C, of parameter u, how many tangents to C (other than the tangent at M) go through M? The parameter t of the contact point of such a tangent is a solution of $ut^2 = 1$ for a nodal curve, and of $2t + u = k$ for a cuspidal curve. In the first case there are generally two solutions, but only one in characteristic two. In the second case there is one solution in general, and none in characteristic two, unless M is the point common to all tangents.

(3) A *point of inflection* I of a plane curve C is a simple point where the tangent to C intersects C with multiplicity at least three. At a point of inflection, the three points the tangent have in common with a cubic coincide. For a nodal cubic, the parameter t of such a point satisfies $t^3 = 1$, which shows there are, in general, three points of inflection with parameters 1, j and j^2 (which are collinear, since $1 \cdot j \cdot j^2 = 1$); in characteristic three there is only one, with parameter 1. For a cuspidal curve, the equation is $3t = k$, so there is in general only one point of inflection; in characteristic three there are none, unless $k = 0$, when

all points are inflection points (for example, all the tangents to $y = x^3$ are horizontal, since $y' = 3x^2 = 0$, and they intersect the curve with multiplicity three).

(4) By the very nature of the parametric representation ($y = tx$) on a unicursal cubic C, we see that a projective transformation or involution on C is induced on C by a projective transformation or involution on the pencil of lines that go through the singular point O. The particular involutions of the form $tt' =$ constant in the nodal case, or $t + t' =$ constant in the cuspidal case, have a simple geometric translation: two point M and M' of C are homologous if and only if they are collinear with a fixed point F on C (this situation is analogous to that of Frégier's theorem). On the pencil of lines that go through O this means, in the nodal case, that 0 and ∞ are homologous, that is, the two tangents to C at O are homologous; and, in the cuspidal case, the tangent at the cusp (with parameter ∞) is the line fixed by the involution. The reader is encouraged to work out the corresponding statements in characteristic two and three.

(5) Theorem 40 can be easily generalized to the $3d$ intersection points of C with a curve D of degree d. If $G(x, y, z) = 0$ is the homogeneous equation of D, these points correspond to the roots of $G\big(t, t^2, F(1, t)\big) = 0$ in the nodal case, and $G\big(1, t, F(1, t)\big) = 0$ in the cuspidal case (with notations as above). The terms in t^{3d} and t^0 in the nodal case, and those in t^{3d} and t^{3d-1} in the cuspidal case, derive only from the term in z^d of G. Thus the product or sum of the roots of the equation does not depend on D. If we take for D the union of d lines, we see that the parameters t_1 of the $3d$ intersections of C with an arbitrary curve D of degree d satisfy

$$\begin{aligned} t_1 t_2 \ldots t_{3d} &= 1 \qquad &&\text{if } C \text{ is nodal,} \\ t_1 + t_2 + \cdots + t_{3d} &= dk \qquad &&\text{if } C \text{ is cuspidal.} \end{aligned} \tag{56}$$

The converse can be shown by uniqueness, if we remark that there exists at least one curve of degree d containing $3d - 1$ given points, because the number of possible coefficients is $\frac{1}{2}(d+1)(d+2)$, which is more than $3d$. Thus the curves of degree d going through $3d - 2$ fixed and one variable point of C determine on C an involution of the type described in (4) above.

It follows from (56) that if C is nodal, a conic D is tritangent to C at the points with parameters t, t' and t'' if and only if $t^2 t'^2 t''^2$, that is, $tt't'' = 1$ or $tt't'' = -1$. In the first case the three points are collinear and D is just twice the line containing the points. In the second case (except in characteristic 2) the conic is non-degenerate; the contact points are the conjugates of three collinear points, where we call two points $M, M' \in C$ *conjugate* if the tangents to C at M and M' intersect on C. (By (2), this means that their parameters t and t' satisfy $t + t' = 0$.)

The corresponding result in the cuspidal case is left to the reader.

(6) A curve C of the Euclidean plane is *circular* if it goes through the cyclic points (section 1.6). Let c and $\bar{c}$ be the parameters of the cyclic points on C. By (56), four points of C, with parameters t_1, t_2, t_3, t_4, are cocircular if and only if

$$t_1t_2t_3t_4c\bar{c} = 1 \qquad \text{if } C \text{ is nodal,}$$
$$t_1 + t_2 + t_3 + t_4 + c + \bar{c} = 2k \qquad \text{if } C \text{ is cuspidal.}$$

Notice that $1/c\bar{c}$ or $k - c - \bar{c}$, respectively, is the parameter of the third point at infinity of C, which is real. Although the parameter t is the slope y/x, one cannot say that $c = i$ unless the x- and y-axes are perpendicular. It is always possible to choose them to be so when C is cuspidal, the only constraint being that the tangent at the cusp is the y-axis. But, in the nodal case, the axes are the tangents to C at the origin, and there is no reason why they should be perpendicular. When they are, C is a *strophoid*.

The osculating circle at a point of parameter t of a nodal (resp. cuspidal) circular cubic C intersects C again at the point with parameter $u = 1/t^3c\bar{c}$ (resp. $u = 2k - c - \bar{c} - 3t$). A short calculation, as in (1), shows that the osculating circles to C at four cocircular points intersect C in four cocircular points.

(7) Let C be a strophoid. Since the parameters i and $-i$ of the cyclic points are the negative of one another, the points are conjugate in the sense of (5): the tangents at these points intersect at a point $F \in C$, called the *singular focus* of C. Let j be the involution determined on C by the lines that go through F; the cyclic points are left fixed by j (cf. (4)). Thus the involution given by j on the pencil of lines through the origin leaves fixed the isotropic lines, and is an orthogonality involution (theorem 32). If M and M' are points of C collinear with F, the lines OM and OM' (where O is the origin) are orthogonal.

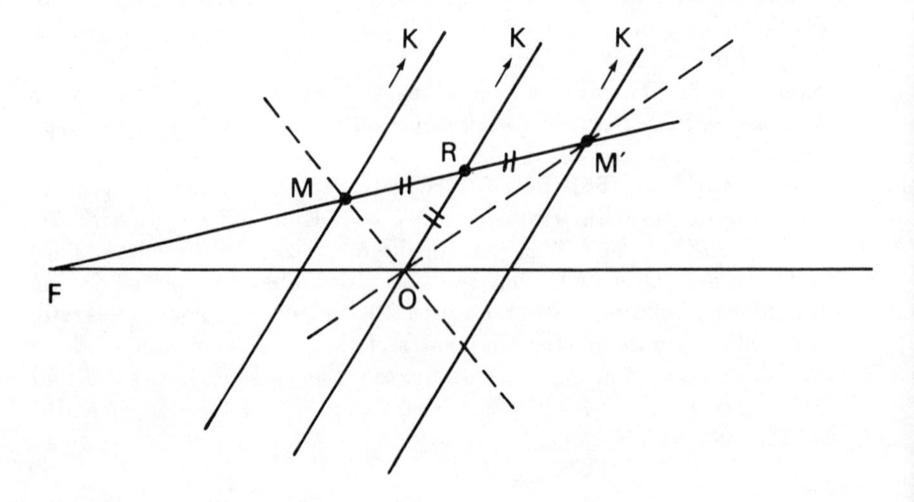

The parameter of F is $1/i^2 = 1/(-i)^2 = -1$; thus the parameters t, t' of M, M' satisfy $tt' = -1$.

On the other hand the real point at infinity C, say K, has parameter 1, since $i(-i)1 = 1$. The line KM intersects C again at a point P with parameter $u = 1/t$, where t is the parameter at M; if M' and P' are points with parameters t' and u' such that $tt' = uu' = -1$, M' and P' are collinear with K. Thus the line KM uniquely determines another line KM', and the correspondence is rational, hence an involution. The fixed lines of this involution are KO and the line at infinity. Thus the point $R = KO \cap MM'$ is the harmonic conjugate of the point at infinity with respect to M and M' (theorem 30); in other words, R is the midpoint of M and M'. Since MOM' is a right angle, the segments RM, RM' and RO have the same length. This gives an elementary definition for the strophoid: given two points O and F and a line $D \ni O$ (which plays the role of OK), mark on every line going through F two points M and M' such that $RM = RM' = RO$, where R is the point where the line meets D. The points M and M' describe the strophoid.

Its tangents at the double point O are the bisectors of D and OF. If D and OF are orthogonal, the strophoid is *right*.

(8) A cuspidal circular cubic is called a *cissoid*. Its equation is of the form $(x^2 + y^2)(ax + by) - x^2 = 0$. Since the cyclic points have parameter values i and $-i$, the real point at infinity has parameter value k (where k is given by the collinearity condition, theorem 40). A parallel D to the real asymptote of C meets C at two points M, M' with parameter values t, t' such that $t + t' + k = k$, that is, $t + t' = 0$; thus the lines OM and OM' are symmetric with respect to the y-axis (which is the tangent at the cusp). If D is tangent to C, we have $t = t' = 0$ and the contact point is on the line orthogonal to the tangent at the cusp and containing the cusp, that is, the x-axis.

(9) Let C be an arbitrary unicursal cubic. If two lines D and D' intersect C at a, b, c and a', b', c', respectively, the points a'', b'', c'' where $D_{aa'}, D_{bb'}, D_{cc'}$ intersect C are collinear. For (identifying points with their parameters) we have, in the nodal case, $abc = 1$, $a'b'c' = 1$, $a'' = 1/aa'$, $b'' = 1/bb'$, $c'' = 1/cc'$, whence $a''b''c'' = 1$; the cuspidal case is similar.

2.7. The Complex Projective Line and the Circular Group

The complex projective line $\mathbf{P}_1(\mathbf{C}) = \hat{\mathbf{C}}$ is obtained by adjoining to the complex plane $\mathbf{C}$ a point at infintiy, denoted by ∞.

The resulting object is not the same as the real projective plane, obtained by adjoining to the ordinary plane a whole line of points at infinity. $\mathbf{P}_1(\mathbf{C})$ and $\mathbf{P}_2(\mathbf{R})$ are not even equivalent in the topological sense (that is, homeomorphic), since the former is orientable but not the latter.

The construction of $\hat{\mathbf{C}}$ can be visualized by means of the *stereographic projection* of the unit sphere $S = S^2 = \{(x, y, z) \mid x^2 + y^2 + z^2 = 1\}$. The line joining the north pole $N = (0, 0, 1)$ to the point $(u, v, 0)$ on the equatorial plane (the xy-plane) meets S again at

$$(57) \qquad (x, y, z) = \frac{1}{u^2 + v^2 + 1}(2u, 2v, u^2 + v^2 - 1).$$

Conversely, given a point $(x, y, z) \in S$ distinct from N, the line joining it with N intersects the equatorial plane at the point

$$(58) \qquad (u, v) = \frac{1}{1 - z}(x, y).$$

Thus we get a bijection (in fact, a homeomorphism) between the equatorial plane, which can be identified with $\mathbf{C}$, and the sphere S minus its north pole. It follows that $\hat{\mathbf{C}}$ can be identified with the sphere S.

Formulas (57) and (58) show that S is a rational surface in the sense of the preceding section.

Circles in S are the intersections of S with planes $ax + by + cz + d = 0$. By (57), the image on the plane of such a circle has equation

$$(59) \qquad (c + d)(u^2 + v^2) + 2au + 2bv + (d - c) = 0.$$

This is the equation of a circle, unless $c + d = 0$, in which case the plane, hence the circle on the sphere, contains the north pole. The plane image of this kind of circle is a line. Conversely, by appropriately choosing coefficients, we see that every circle and every line on the equatorial plane can be obtained in this way.

To simplify statements we introduce a catch-all definition: a *circle-line* on $\hat{\mathbf{C}}$ is a circle on S. The trace of a circle-line on $\mathbf{C}$ is a circle or a line, the latter when the circle-line goes through the point at infinity.

Theorem 41. *Four points $a, b, c, d \in \hat{\mathbf{C}}$ are cocircular or collinear if and only if their cross-ratio (a, b, c, d) is real.*

Proof. Recall that $(a, b, c, d) = (c - a)(d - b)/(c - b)(d - a)$ (theorem 21, section 1). The argument V of $(c - a)/(c - b)$ is the angle between the vectors $(c - b)$ and $(c - a)$; similarly, the argument W of $(d - a)/(d - b)$ is the angle between $(d - b)$ and $(d - a)$. Thus the cross-ratio (a, b, c, d) is real if and only if the angle $V - W$ equals 0 or π. It amounts to the same to say that the angles (between lines) $(\widehat{D_{bc}, D_{ac}})$ and $(\widehat{D_{bd}, D_{ad}})$ are the same; but it's well-known from elementary geometry that this is equivalent to a, b, c, d being cocircular or collinear. $\square$

We define the *Möbius group* $\mathbf{G}$ as the group of transformations of $\hat{\mathbf{C}}$ of the form $h(z) = (az+b)/(cz+d)$ or $h(z) = (a\bar{z}+b)/(c\bar{z}+d)$. Its elements, or in some texts just those elements of the form $h(z) = (az+b)/(cz+d)$ (the projective transformations) are called *Möbius transformations*. The projective group $\mathrm{PGL}(\mathbf{C}^2) = \mathbf{G}^+$ is an index-two subgroup of $\mathbf{G}$; its complement is denoted by $\mathbf{G}^-$. By theorem 20 in section 1 and formula (45) in section 3, Möbius transformations preserve or conjugate cross-ratios, according to whether they belong to $\mathbf{G}^+$ or $\mathbf{G}^-$. In particular, by theorem 41, the Möbius group preserves circle-lines. We'll see a converse below.

Examples of Möbius transformations

(1) The Möbius transformations of the form $s(z) = az+b$ or $t(z) = a\bar{z}+b$ are those that leave fixed the point at infinity. Since they are compositions of translations $f(z) = z+b$, rotations about the origin $r(z) = uz$ (where $|u| = 1$), homotheties centered at the origin $h(z) = kz$ (where $k \in \mathbf{R}^*$) and the reflection through the real axis $g(z) = \bar{z}$, we conclude that they are similarities of $\mathbf{C}$. Conversely, every similarity of the Euclidean plane is of this form, as follows: by composing with a homothety and a translation, we reduce to an isometry fixing the origin; by elementary geometry, such an isometry is an $\mathbf{R}$-linear transformation whose matrix, in an orthonormal basis, is of the form $\begin{pmatrix} a & -b \\ b & a \end{pmatrix}$ (rotation) or $\begin{pmatrix} a & b \\ b & -a \end{pmatrix}$ (reflection through a line), with $a^2+b^2 = 1$. (Verifying of this fact is particularly simple if we have defined $\mathbf{C}$ as the set of real matrices of the form $\begin{pmatrix} a & -b \\ b & a \end{pmatrix}$.)

We recall that this use of complex numbers is very convenient in the study of elementary geometric transformations. It shows, for example, that an orientation-preserving similarity $s(z) = az+b$ that is not a translation ($a \neq 1$) has a unique fixed point. An orientation-reversing similarity $t(z) = a\bar{z}+b$ that is not an isometry ($a\bar{a} \neq 1$) is also easily shown to have a unique fixed point. In general, an orientation-reversing isometry has no fixed point, being the composition of a reflection with a glide along the reflection axis; if there is a fixed point, there is a whole line D of them, and the isometry is the reflection through D.

(2) The inversion with pole at the origin and power $k \in \mathbf{R}^*$ (section 1.6) is described by $z' = k/\bar{z}$. This follows from formula (25) or from the fact that z and $1/\bar{z}$ have the same argument and $|z| \cdot |z'| = k$. The inversion with pole a and power k is described by

$$z' - a = \frac{k}{\overline{z-a}} = \frac{k}{\bar{z}-\bar{a}}. \tag{60}$$

This is clearly an involution in $\mathbf{G}^-$. The next theorem provides a converse.

Theorem 42. *An involutive Möbius transformation f that is not a projective transformation is either an inversion or the reflection through a line.*

Proof. If the point at infinity is fixed, we have an orientation-reversing similarity $f(z) = a\bar{z} + b$; since $z = f^2(z) = a\bar{a}z + a\bar{b} + b$, we have $a\bar{a} = 1$ and f is an isometry. If c is a point not fixed by f, the midpoint of c and $f(c)$ is fixed as is the line D equidistant to these two points; this implies that f is the reflection through D.

If ∞ is not fixed, let $m = f(\infty)$. Since $f(z) - m$ equals ∞ for $z = m$ and 0 for $z = \infty$, it is of the form $d/(\bar{z} - \bar{m})$. Setting $f(z) = t$, we have $(t - m)(\bar{z} - \bar{m}) = d$. Switching z and t we see that $d = \bar{d}$, hence $d \in \mathbf{R}$. Thus f is an inversion (of center m and power d). □

Remark on fixed points. We know that a projective transformation (distinct from the identity) has either one or two fixed points (section 4). An inversion has a whole circle of fixed points if its power is positive, and no fixed points if it is negative.

If $f \in \mathbf{G}^-$ has at least one fixed point, jfj^{-1}, where j is some projective transformation, leaves ∞ fixed; this brings us back to the case of an orientation-reversing similarity. But we have seen that such a similarity has either one finite fixed point (the general case), none (reflection-cum-glide), or a whole line of them (reflection). The only case when the similarity is involutive is the last. Returning to f (which is involutive if and only if jfj^{-1} is), we see that the following cases can occur:

- no fixed point;
- a single fixed point (case of a conjugate of a reflection-plus-glide)
- two fixed points (the general case among anti-projective transformations having at least one fixed point)
- a circle-line of fixed points (case of a reflection or an inversion with positive power).

Theorem 43. *A bijection $f : \hat{\mathbf{C}} \to \hat{\mathbf{C}}$ transforms circle-lines into circle-lines if and only if it is a Möbius transformation.*

Proof. Sufficiency has been seen. For necessity we can assume, after composing f with a projective transformation if necessary, that the point at infinity is fixed under f. Then f transforms lines into lines and circles into circles ("lines" and "circles" being used in the restricted sense). In other words, f preserves everything that can be built with ruler and compass. And we will see that such is the case with harmonic quadrilaterals (quadruples of points a, b, c, d such that $(a, b, c, d) = -1$).

After we have shown that harmonic quadrilaterals are preserved, theorem 28 (section 3) implies that f is the composition of a projective transformation with an automorphism s of $\mathbf{C}$. Since f takes cocircular points into cocircular points, theorem 41 shows that s transforms every real number into a real number. Thus s induces the identity on $\mathbf{R}$ (cf. corollary to theorem 28) and must be the identity or the conjugation map on $\mathbf{C}$.

There remains to show that harmonic conjugates can be found with ruler and compass. We will need the following lemma:

Lemma. *Let a, b, c, d be four points on a circle-line D. The cross-ratio $(a, b, c, d)_{\mathbf{C}}$, computed in $\hat{\mathbf{C}}$, is equal to the cross-ratio $(a, b, c, d)_D$, computed in D (which, being a line or a conic, has a standard real projective line structure).*

Proof. If D is a line, it has a parametric representation $z = pt + q$, where $p, q \in \mathbf{C}$ and t is a real parameter; this immediately implies the lemma. If D is a circle, we apply an inversion j with pole $m \in D$. By definition, the cross-ratio $(a, b, c, d)_D$ is the cross-ratio of the lines ma, mb, mc, md (theorem 33 in section 5), which is the same as the cross-ratio $\big(j(a), j(b), j(c), j(d)\big)_{j(D)}$. By the first part of the proof, this is identical with $\big(j(a), j(b), j(c), j(d)\big)_{\mathbf{C}}$, hence with the conjugate of $(a, b, c, d)_{\mathbf{C}}$ because $j \in \mathbf{G}^-$. The lemma follows since $(a, b, c, d)_{\mathbf{C}}$ is real (theorem 41). □

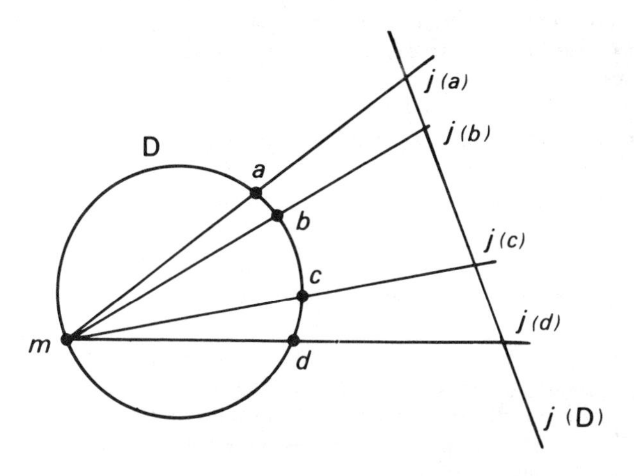

Remark. The two cross-ratios are not longer equal when D is a conic other than a circle. The cross-ratio in $\mathbf{C}$ is not even real.

Now if four points form a harmonic quadrilateral, $(a, b, c, d) = -1$, they lie on the same circle-line D, and, by the lemma, c and d are homologous under the involution (on D) having fixed points a and b (theorem 30 in section 4). When D is a circle, this means that c and d are collinear with the Frégier point of the involution (theorem 35 in section 5), which is the intersection of the tangents to D at a and b. Thus the fourth point d can be obtained with ruler and compass from a, b and c, and this construction can be extended to $f(a), f(b), f(c), f(d)$. If D is a line, we've seen a construction for d with ruler only in section 3. □

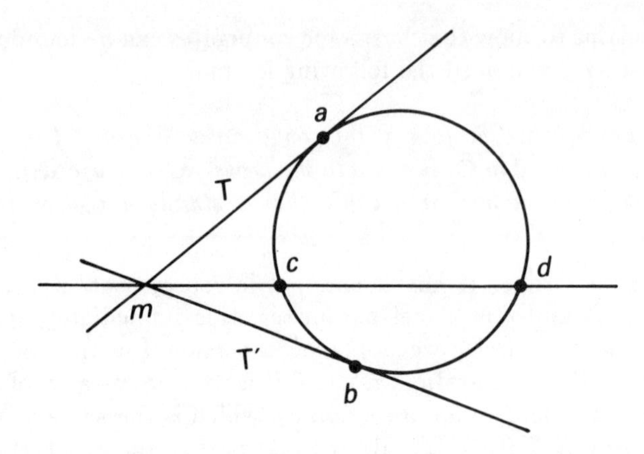

For those who prefer a more formal proof, here it is. Let a, b, c, d be such that $(a, b, c, d) = -1$. Assume, for concreteness, that they belong to the same circle D. By assumption, $f(a), f(b), f(c), f(d) \in f(D)$, which is also a circle. If T and T' are the tangents to D at a and b, respectively, $f(T)$ and $f(T')$, which are lines, must be the tangents to $f(D)$ at $f(a)$ and $f(b)$, respectively. Let m be the intersection of T and T' (possibly at infinity); then $f(m)$ is the intersection of $f(T)$ and $f(T')$. Since m, c and d are collinear by Frégier's theorem, the same is true about $f(m)$, $f(c)$ and $f(d)$. Thus $\big(f(a), f(b), f(c), f(d)\big) = -1$.

Remark. The preceding argument gives a description for any involution j of $\hat{\mathbf{C}}$ having two fixed points a and b: for every $m \in \hat{\mathbf{C}}$ outside the line ab, draw the circle D going through a, b, m, then its tangents T, T' at a, b. They intersect at some point p; then $j(m)$ is the second intersection of pm with D.

Yet another proof for theorem 43: we can assume that f takes lines into lines and that $f(0) = 0$. By the remark following theorem 7 (section 1.3), f is a semi-linear bijection of the $\mathbf{R}$-vector space $\mathbf{C}$. But every automorphism of $\mathbf{R}$ is trivial, so f is $\mathbf{R}$-linear. Since it preserves circles, it multiplies the form $x^2 + y^2$ by a constant; thus it must be a similarity.

Theorem 44.

(a) *Every $f \in \mathbf{G}^-$ is the product of at most three inversions or reflections.*

(b) *Every $g \in \mathbf{G}^+$ is the product of at most four inversions or reflections.*

(c) *These results are best possible.*

Proof. We prove (a). By conjugation—which preserves the property of being involutive—we can assume that $f(\infty)$ is a finite number a. Let i be an inversion with pole a. Then $if(\infty) = \infty$ and if is an orientation-preserving similarity. Let b be its ratio and h the homothety of center a and ratio $1/b$. Then hif is an orientation-preserving isometry, rotation or translation, which is known to be the product ss' of two reflections. On the other hand hi is an inversion j of pole a, so $jf = ss'$ and $f = jss'$.

By composing g with an inversion or reflection we reduce (b) to (a).

We need an odd number of inversions or reflections to make up an element of $\mathbf{G}^-$, so (a) is best possible because there are elements of $\mathbf{G}^-$ that are neither inversions nor reflections (e.g., $z' = 3\bar{z}$). To complete the proof of (c), we must show that there are projective transformations that are not the composition of two inversions or reflections. Let $h(z) = az$ be a similarity that is neither a homothety nor a rotation (e.g., $h(z) = 23iz$). If h were the product of two inversions, they would have the same pole (since $h(\infty) = \infty$) and h would be a homothety, which is against the assumptions. If h were the product of two reflections, it would be an isometry, which again we have excluded. Finally, if h were the product of an inversion and a reflection, we'd have $h(\infty) \neq \infty$, also against the assumptions. □

In fact, a projective transformation that is the product of two inversions or symmetries is conjugate to an isometry or a homothety.

2.8. Topology of Projective Spaces

We will assume here that the field of scalars K is *locally compact and non-discrete*. We recall (see [Bo1, chapter VI], for example) that these fields are: (1) $\mathbf{R}$ and $\mathbf{C}$; (2) the finite extensions of the p-adic fields $\mathbf{Q}_p$, for p prime ($\mathbf{Q}_p$ is the completion of $\mathbf{Q}$ for the p-adic valuation); and (3) the fields $\mathbf{F}((X))$ of formal power series in one variable over a finite field $\mathbf{F}$. A finite-dimensional topological vector space E over such a field K must have the product topology, that of K^n (see [Bo2, chapter I]); thus it is locally compact. Such a field K has a valuation v, and E has a norm N. We will assume that, for every $x \in E$, there exists $a \in K$ such that $N(x) = v(a)$.

This is automatically true if $K = \mathbf{R}$ or $\mathbf{C}$: take $a = N(x)$, which is real. Otherwise one can choose N to satisfy the condition (take a "sup" norm).

This condition means that every vector line of E intersects the *unit sphere* S, defined by $N(x) = 1$.

Theorem 45. *With the hypotheses and notations above, the projective space $\mathbf{P}(E)$, endowed with the quotient topology induced from $E \setminus \{0\}$, is Hausdorff and compact. It is also the quotient S/U of the unit sphere S by the subgroup U of elements of K having absolute value 1 (which acts on S by homotheties).*

Proof. Recall (see [Bo3, chapter I]) that the quotient of a topological space V by an equivalence relation R is Hausdorff if and only if the saturation of every open set U is open and the graph of the relation R in $V \times V$ is closed.

Here, where V is $E\setminus\{0\}$ and R is collinearity, the saturation of an open set U is the union of its homothetic images aU, for $a \in K^*$. This is a union of open sets, hence open. For the second condition, take coordinates $(x_1,\dots,x_n)$ in E and corresponding coordinates $(y_1,\dots,y_n)$ in the second factor of $E \times E$; the graph of R is defined by the polynomial equations $x_i y_j - x_j y_i = 0$, for $i \neq j$, and is consequently closed. Thus $\mathbf{P}(E)$ is Hausdorff.

The canonical map $p : E\setminus\{0\} \to \mathbf{P}(E)$ is continuous by definition, and we have $p(S) = \mathbf{P}(E)$. Thus $\mathbf{P}(E)$ is compact, being the continuous image of a compact in a Hausdorff space. It is also the quotient of the sphere S by the equivalence relation induced by R on S; now, if $N(x) = N(y) = 1$, we have a relation of the form $x = ay$, for $a \in K^*$, if and only if $v(a) = 1$. This means that $\mathbf{P}(E) = S/U$. □

Examples of real projective spaces

For $K = \mathbf{R}$, the group U reduces to $\{1, -1\}$, so $\mathbf{P}_n(\mathbf{R})$ is the quotient of the unit sphere S^n of $\mathbf{R}^{n+1}$ by the antipodal relation, which identifies pairs of diametrically opposed points in S^n.

For $n = 1$, S^1 is the same as the group of complex numbers of absolute value 1. This group has an endomorphism h, given by $h(z) = z^2$, whose kernel is exactly U. Thus $\mathbf{P}_1(\mathbf{R})$ is the same as the image of h, which is the circle S^1.

For $n = 2$, we obtain $\mathbf{P}_2(\mathbf{R})$ by taking the northern hemisphere of the unit sphere S^2 and identifying pairs of diametrically opposed points on the equator. This identification is hard to visualize; in fact, it can be shown that it is impossible to embed $\mathbf{P}_2(\mathbf{R})$ as a surface in $\mathbf{R}^3$. The real projective plane is a non-orientable surface: take, at a point a of the equator, a basis for the tangent space to S^2, where the x-axis points east and the y-axis north; then move a along the equator until it reaches the diametrically opposed point a'. The new basis obtained by this movement has the x-axis still pointing east and the y-axis north; but the basis at a obtained from this one by the antipodal map has its x-axis pointing east and the y-axis pointing south, so its orientation is opposite the orientation of the original basis.

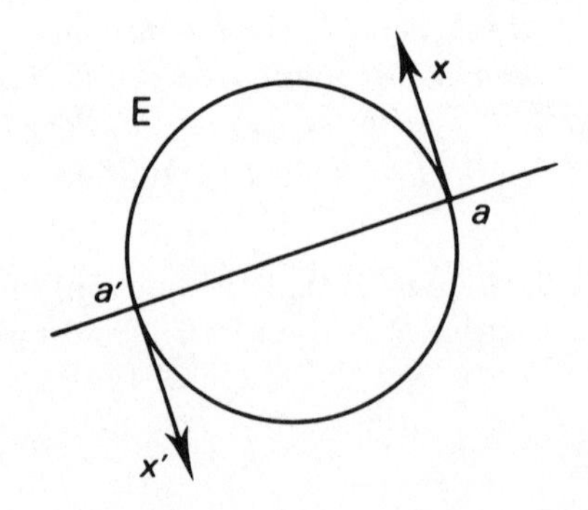

The best-known example of a non-orientable surface is the *Möbius strip*. The projective plane cannot be the same as the Möbius strip, because it doesn't have a boundary. But there are Möbius strips contained in $\mathbf{P}_2(\mathbf{R})$. To see this, take two points a and b close to one another and lying on the equator, and consider the planes through a and b that are perpendicular to the line ab. These planes determine on the northern hemisphere a strip B, whose ends are the arcs ab and $a'b'$ on the equator, where a' and b' are the points diametrically opposed to a and b. The identification of ab with $a'b'$ gives the twist to the strip.

The lines of $\mathbf{P}_2(\mathbf{R})$ are the images of great circles, the intersections of the sphere with planes going through the origin.

The sphere S^3 has a (non-abelian) group structure induced by quaternion multiplication. For this reason the projective space $\mathbf{P}_3(\mathbf{R})$ can be identified with the group $S^3/\{1,-1\}$. Being a topological group, this projective space is orientable: we can obtain a consistent orientation at every point by applying left translations to a tangent basis at 1.

The same reasoning implies that S^2 cannot be given a "reasonable" group structure.

One can show that $\{1,-1\}$ is the center of S^3.

Examples of complex projective spaces

For $K = \mathbf{C}$, the group U of theorem 45 is the group of complex numbers of absolute value 1, which is homeomorphic to S^1. The unit sphere S of $\mathbf{C}^n$, defined by the equation

$$z_1\bar{z}_1 + z_2\bar{z}_2 + \cdots + z_n\bar{z}_n = 1,$$

has real dimension $2n-1$.

Thus $\mathbf{P}_1(\mathbf{C})$ is the space of orbits of the action of U on S^3. These orbits are circles, and the set of them is the so-called *Hopf fibration* on S^3. We saw at the beginning of section 7 that $\mathbf{P}_1(\mathbf{C})$ is homeomorphic to S^2. The base of the Hopf fibration (that is, the quotient space), must then be S^2. We say that S^3 is a circle bundle over S^2.

By choosing a copy of $\mathbf{C}$ inside the quaternions, we can see U as a subgroup of S^3. But it is not a normal subgroup, so the set S^3/U of cosets Ux is not a group.

More generally, S^{2n+1} is a circle bundle over $\mathbf{P}_n(\mathbf{C})$.

CHAPTER 3

Classification of Conics and Quadrics

A hypersuface in an affine or projective space defined by a quadratic equation is called a *hyperquadric*, or simply a *quadric*; in dimension two quadrics are called *conics*.

Classifying the quadrics in a space E means listing their orbits under the action of an appropriate group G of automorphisms of E: the projective group if E is a projective space; the affine group if E is a affine space; and the group of isometries (or similarities) if E is a Euclidean space over $\mathbf{R}$.

To do this we can, for example, exhibit one element for each orbit. Since G acts simply transitively on the frames of E, we can, to find the orbit of a quadric S, take a frame R intrisically attached to S; the orbit of S is then formed of all quadrics having, in other frames of E, the same equation that S has in R.

The elements of G which stabilize the quadric S form a subgroup $G(S)$ of G and the orbit of S is in bijection with the set of classes $G/G(S)$. Thus bigger $G(S)$ is, the smaller the orbit.

3.1. What Is a Quadric?

We saw in section 1.8 that over an algebraically closed field K, the equation of a quadric is determined, up to a scalar, by the set of its points. This is not true over other fields: the set of *rational points*—those having affine or homogeneous coordinates in K—of a quadric S does not necessarily determine its equation.

For example, in $\mathbf{R}^2$, the equations $x^2+y^2+1=0$ and $x^2+xy+y^2+3=0$ define the same (empty) set. In $\mathbf{R}^3$, the equations $x^2+y^2=0$ and

$3x^2 + 5y^2 = 0$ both define the z-axis ($x = y = 0$); similarly $x^2 - 2y^2 = 0$ and $3x^2 - y^2 = 0$ in $\mathbf{F}_5^3$.

We do have the following result:

Theorem 46. *Let $F(x) = 0$ be an equation for a quadric S in an affine or projective space over a field K. If the equation is satisfied for a simple point of S, and if $K \neq \mathbf{F}_2$ in the affine case, every other equation of S is proportional to F.*

Proof. Let's start with the affine case, and take the given simple point as the origin. The equation $F(x) = 0$ can then be written $Q(x) + L(x) = 0$, where Q is a quadratic form and L is a linear form in $x = (x_1, \ldots, x_n)$ with $L \neq 0$. Let $Q_0(x) + L_0(x) = 0$ be another equation defining the same subset $S \in K^n$. We will need the following lemma:

Lemma. *Let $P(x_1, \ldots, x_n)$ be a homogeneous quadratic polynomial over a field K.*

(a) *If $P(a) = 0$ for every $a = (a_1, \ldots, a_n) \in K^n$, then $P = 0$.*
(b) *If $P(a) = 0$ for every $a = (a_1, \ldots, a_n) \in K^n$ such that $a_1 \neq 0$ and if $K \neq \mathbf{F}_2$, then $P = 0$.*
(c) *If $P(0, a_2, \ldots, a_n) = 0$ for every $a = (a_2, \ldots, a_n) \in K^{n-1}$, then P is a multiple of x_1.*

Proof. Write $P(x_1, \ldots, x_n) = cx_1^2 + x_1 P_1(x_2, \ldots, x_n) + P_2(x_2, \ldots, x_n)$, where P_i is homogeneous of degree i. To show (a) we use induction on n, starting with the trivial case $n = 1$. Taking $a_1 = 0$ and $a_2, \ldots, a_n$ arbitrary, the induction hypothesis gives $P_2 = 0$; this gives us (c) in the bargain. Taking $a_1 = 1$, $a_2 = \cdots = a_n = 0$, we get $c = 0$. Finally, for $a_1 = 1$ and $a_2, \ldots, a_n$ arbitrary, we get $P_1(a_2, \ldots, a_n) = 0$, which implies $P_1 = 0$ because P_1 is linear.

We now show (b). The relation $P(1, 0, \ldots, 0) = 0$ gives $c = 0$. If $P_1 = 0$, then $P = P_2$, which is zero by (a). Otherwise, let $b \in K^{n-1}$ be such that $P_1(b) \neq 0$; then $a_1 P_1(b) + P_2(b) = 0$, that is, $a_1 = -P_2(b)/P_1(b)$ for every $a_1 \neq 0$ in K, which is impossible unless $K = \mathbf{F}_2$. □

Over $\mathbf{F}_2$ the property in (b) does not hold for $xy + y^2$.

We continue with the proof of theorem 46, in the affine case. Intersect S with lines $x' = tx$, for $t \in K$, as if to parametrize S. The equation $0 = Q(tx) + L(tx) = t^2 Q(x) + tL(x)$ (in t) has as roots, in addition to $t = 0$, the root of

$$tQ(x) + L(x) = 0. \tag{E}$$

Thus (E) and the corresponding equation (E_0) involving Q_0 and L_0 have the same roots for every $x \in K^n$.

If $L(x) = 0$ and $Q(x) \neq 0$, this root is $t = 0$, so $L_0(x) = 0$. If $L(x) = 0$ and $Q(x) = 0$, equation (E) is always satisfied, hence so is (E_0), and we conclude that $Q_0(x) = L_0(x) = 0$. In any case $L(x) = 0$ implies $L_0(x) = 0$, so $\ker L \subset \ker L_0$, and L_0, being linear, is of the form $L_0 = cL$ with $c \in K$.

Finally, the determinant $Q(x)L_0(x) - L(x)Q_0(x)$ of (E) and (E_0) is always zero (otherwise the solution $(t, 1)$ of the system would be $(0,0)$). Since $L_0 = cL$, this implies $\bigl(Q_0(x) - cQ(x)\bigr)L(x) = 0$ for every $x \in K^n$. Applying part (b) of the lemma, we get $Q_0 = cQ$, whence $Q_0 + L_0 = c(Q + L)$, completing the proof of the affine case.

The projective statement is weaker than the affine unless $K = \mathbf{F}_2$. In this case, let $G(x) = 0$ and $G_0(x) = 0$ be two homogeneous equations for the same set S. Since $G(x) \neq 0$ here means $G(x) = 1$, and similarly for G_0, the forms G and G_0 have the same value everywhere, and are thus equal by part (a) of the lemma. □

The affine conics $x^2 + xy + y^2 + x = 0$ and $xy + y^2 = 0$ over $\mathbf{F}_2$ have the same rational points $(0,0)$, $(1,0)$, $(1,1)$. But the first has no rational point at infinity, whereas the second has two.

The existence of a simple rational point was not used when $K = \mathbf{F}_2$ in the projective case.

3.2. Classification of Affine and Euclidean Quadrics

We will consider here quadrics in an n-dimensional affine space E over a field K of characteristic $\neq 2$. This will allow us to use the decomposition of quadratic forms into squares. As we go along we will add, where appropriate, extra results for two important cases: $K = \mathbf{R}$, and E a real Euclidean space.

Once we have chosen an origin for E, the equation of a quadric S can be written $Q(x) + L(x) + k = 0$, where Q is a quadratic form, L a linear form and k a constant. Moving the origin to a transforms this equation into $Q(x+a) + L(x+a) + k = 0$, or, denoting by B the bilinear form associated with Q,

$$Q(x) + \bigl(2B(x,a) + L(x)\bigr) + Q(a) + L(a) + k = 0.$$

Thus Q is independent of the choice of an origin, but $L(x)$ gets replaced by

$$L'(x) = L(x) + 2B(x,a). \tag{61}$$

Denote by $r \leq n$ the rank of Q, and let $(e_1, \ldots, e_n)$ be an orthogonal basis for Q. If E is Euclidean, we assume that $(e_1, \ldots, e_n)$ is also orthonormal with respect to the Euclidean inner product (x, y).

We recall how to show that such a basis exists. Since the inner product is non-degenerate, there exists, for each $x \in E$, a unique linear map $s : E \to E$ such that $B(x,y) = \bigl(s(x),y\bigr)$. The symmetry of B means that $\bigl(s(x),y\bigr) = \bigl(x,s(y)\bigr)$. We deduce that if a subspace F of E is left invariant by s, so is its orthogonal complement $F^\perp$ with respect to the inner product (take $x \in F^\perp$ and $y \in F$). Let a an eigenvalue, possibly complex, of s, and z a corresponding eigenvector in the complexification of E: $s(z) = az$. Then $s(\bar z) = \overline{az}$, so $\bigl(s(z),\bar z\bigr) = \bigl(z,s(\bar z)\bigr)$, that is, $a(z,\bar z) = \bar a(z,\bar z)$. If a were not real, we'd have $(z,\bar z) = 0$, whence, setting $z = x+iy$ with $x,y \in E$, $(x,x)+(y,y) = 0$, implying $x = y = 0$ and $z = 0$, contradiction. Thus a and z are real. Normalize z and take it as the first basis vector. Then repeat the procedure with the orthogonal complement $(\mathbf{R}z)^\perp$ of $\mathbf{R}z$; the result follows by recurrence.

Let the equation of S in this basis be

$$a_1x_1^2 + \cdots + a_rx_r^2 + b_1x_1 + \cdots + b_nx_n + k = 0,$$

where $a_1,\ldots,a_r$ are non-zero. The subspace N spanned by $e_{r+1},\ldots,e_n$ is the kernel N of the form Q, that is, it is the space of vectors Q-orthogonal to E. In addition, if E is Euclidean, $\mathbf{R}e_1 + \cdots + \mathbf{R}e_r$ is the orthogonal complement $N^\perp$ of N for the inner product.

After translating the origin, if necessary (replacing x_i, for $i = 1,2,\ldots,r$, by $x_i + b_i/2a_i$), we obtain

$$a_1x_1^2 + \cdots + a_rx_r^2 + b_{r+1}x_{r+1} + \cdots + b_nx_n + k = 0.$$

There are two possible cases, depending on whether or not all the b_1 are zero:

(a) If they are, the equation takes the form

$$a_1x_1^2 + \cdots + a_rx_r^2 + k = 0. \tag{62}$$

This is certainly the case if $r = n$. Intrinsically, this means that $N \subset \ker L$; this condition is independent of the choice of an origin by (61).

(b) If the b_i are not all zero, we choose a basis for N such that the form $b_{r+1}x_{r+1} + \cdots + b_nx_n$ is proportional to a single coordinate form, with we still denote by bx_{r+1}. By translation along the x_{r+1}-axis, we can make k disappear. If E is Euclidean, we can still assume that the basis chosen on N is orthonormal.

In this case we have the equation

$$a_1x_1^2 + \cdots + a_rx_r^2 + bx_{r+1} = 0. \tag{63}$$

If $r < n$ in case (a) or $r+1 < n$ in case (b), S is the product of a quadric in K^r or K^{r+1}, respectively, with an affine space of the complementary dimension. We say that S is a *cylinder*.

In case (a), S is symmetric with respect to the origin and to each coordinate axis. If $r = n$, a short calculation shows that the center of symmetry is unique; we say that S is a *central quadric*.

We will see in chapter 4 that the center of S is the pole of the hyperplane at infinity.

Classification over an algebraically closed field

If K is algebraically closed, or, more generally, if every element of K is a square, we can absorb the coefficients a_i and b of (62) and (63) into the x_i. Thus quadrics over such fields are classified as follows (the cases parallel those above):

(a) the rank r of Q, and whether or not $k = 0$ (if $k \neq 0$ we can take $k = 1$);
(b) the rank r of Q.

Affine classification over $\mathbf{R}$

Sylvester's law of inertia says that the number p of coefficients $a_i > 0$ (and consequently the number $q = r - p$ of coefficients $a_i < 0$) is an invariant of the quadratic form. The pair (p, q) is called the *signature* of Q. Since every positive real number is a square, we can replace each positive a_i by 1 and each negative a_i by -1, in (62) and (63). Thus affine quadrics over $\mathbf{R}$ are classified by

(a) the signature (p, q) of Q, and whether or not $k = 0$ (if $k \neq 0$ we can take $k = 1$, possibly after interchanging p and q);
(b) the signature (p, q) of Q (taking $b = 1$).

The classification is still discrete, as over an algebraically closed field.
Quadrics with signature (p, q) and (q, p) are affinely equivalent in case (b), and in case (a) if $k = 0$.

Affine classification of conics in $\mathbf{R}^2$

If Q has rank two, we're in case (a). If $k \neq 0$, we normalize by setting $k = -1$ and obtain three possible equations:

- $-x^2 - y^2 - 1 = 0$, an empty conic, sometimes called the *imaginary ellipse* (IE).
- $x^2 - y^2 - 1 = 0$, a *hyperbola* (H); the lines $y = x$ and $y = -x$ are called its *asymptotes*, the x-axis the *transverse axis* (it meets H at two real points), and the y-axis the *non-transverse axis*.
- $x^2 + y^2 - 1 = 0$, an *ellipse* (E).

If $k = 0$, we obtain two equations:

- $x^2 + y^2 = 0$, two complex conjugate lines going through the origin.
- $x^2 - y^2 = 0$, two real lines going through the origin.

If Q has rank one, case (a) gives the equation

- $x^2 + k = 0$, with $k = 1$, -1 or 0: two complex conjugate parallel lines if $k = 1$, two real lines if $k = -1$ and a double real line if $k = 0$.

In case (b) we have the equation

- $x^2 + y = 0$, a *parabola* (P).

Of these, EI, H, E and P are irreducible.

Euclidean classification of conics in $\mathbf{R}^2$

Here it's no longer possible to combine the coefficients with the variables, because we want to preserve the property that the basis is orthonormal with respect to the Euclidean inner product. For this reason the classification is no longer discrete; it involves (positive) real parameters. Here are the irreducible conics with their parameters:

$EI: (x/a)^2 + (y/b)^2 + 1 = 0.$
$H: (x/a)^2 - (y/b)^2 - 1 = 0.$
$E: (x/a)^2 + (y/b)^2 - 1 = 0.$
$P: x^2 - 2py = 0.$

For EI and E and we can assume $a \geq b$.

Affine classification of quadrics in $\mathbf{R}^3$

If Q has rank three and $k \neq 0$, normalize by setting $k = -1$. The possible signatures are $(+++)$, $(++-)$, $(+--)$ and $(---)$, so we have the following equations:

- $x^2 + y^2 + z^2 - 1 = 0$, an *ellipsoid* (E).
- $x^2 + y^2 - z^2 - 1 = 0$, a *one-sheet hyperboloid* $(H1)$.
- $x^2 - y^2 - z^2 - 1 = 0$, a *two-sheet hyperboloid* $(H2)$.
- $-x^2 - y^2 - z^2 - 1 = 0$, an empty quadric, called an *imaginary ellipsoid* (IE).

Ellipsoids are affinely equivalent to spheres. One-sheet hyperboloids are obtained by turning a hyperbola around its non-transverse axis, and two-sheet hyperboloids by turning a hyperbola around its transverse axis. It can be shown that by rotating a line D around an axis not coplanar with it, one obtains a one-sheet hyperboloid (a "spaghetti bundle").

If $k = 0$, there are only two inequivalent signatures, $(+++)$ and $(++-)$:

- $x^2 + y^2 + z^2 = 0$, an imaginary cone with vertex at the origin.
- $x^2 + y^2 - z^2 = 0$, a real cone with vertex at the origin.

If Q has rank two, we obtain in case (a) the five equations of conics whose quadratic form has rank two, and the corresponding quadrics are cylinders over these conics: the first three types are called *imaginary cylinder*, *hyperbolic cylinder* and *elliptic cylinder*, respectively, and the last two are pairs of complex conjugate or real planes, respectively, containing the z-axis.

In case (b), we normalize equation (63) by setting $b = -1$. There are only two inequivalent signatures, $(++)$ and $(+-)$, and the respective equations are:

- $x^2 + y^2 - z = 0$, an *elliptic paraboloid* (EP).
- $x^2 - y^2 - z = 0$, a *hyperbolic paraboloid* (HP).

The elliptic paraboloid is obtained by rotating the parabola $z = x^2$, $y = 0$ around the z-axis.

If Q has rank one, we obtain in case (a) one equation $x^2 + k = 0$, for $k = 1, -1$ or 0: two complex conjugate or real planes, or a real double plane. In case (b), we have a single equation, $x^2 + y = 0$, giving a *parabolic cylinder*.

The irreducible quadrics are E, $H1$, $H2$, IE, EP, EH, the four types of cylinders and the two types of cones.

Theorem 47. *With the exception of cylinders, cones, and unions of planes, the only quadrics containing real lines are* $H1$ *and* HP.

Proof. We first show that E, $H2$ and PE contain no real lines. Indeed, the x-coordinate of a real point of the quadric must satisfy $|x| \leq 1$ for E, $|x| \geq 1$ for $H2$ and $x \geq 0$ for PE; thus if any of these quadrics contains a line the x-coordinate of the line is a constant c. On the other hand, the section of a quadric of one of these types by a plane of the form $x = c$ contains a line if and only if the quadric is a PE and $x = 0$; but then the section $y^2 + z^2 = 0$ is the union of two complex conjugate lines.

There remains to show that $H1$ and PH contain real lines. To do this, write their equations under the form

$$(x-1)(x+1) = (z-y)(z+y)$$

and

$$(x-y)(x+y) = z,$$

respectively. One checks immediately that lines of the form

$$x - 1 = s(z \pm y), \quad z \mp y = s(x+1)$$

and

$$x \pm y = s, \quad z = s(x \mp y),$$

respectively, are contained in the quadric. Thus one-sheet hyperboloids and hyperbolic paraboloids contain two one-parameter families of lines $\{D_s | s \in \mathbf{R}\}$ and $\{D_s^* | s \in \mathbf{R}\}$. These lines are called their *rectilinear generators.* □

It is easy to see that two generators D_s and D_t or D_s^* and D_t^* belonging to the same family don't intersect, even at infinity. On the other hand, two generators D_s and D_t^*, one from each family, have a unique intersection point: $\big((1+st)/(1-st), (s-t)/(1-st), (s+t)/(1-st)\big)$ for $H1$, and $\big((s+t)/2, (s-t)/2, st\big)$ for HP.

This situation will be generalized in the next section, when we study Segre varieties.

The Euclidean classification of quadrics in $\mathbf{R}^3$ refines the affine classification by introducing real coefficients into the equations.

3.3. Projective Classification of Real Quadrics

A quadric S of $\mathbf{P}_n(\mathbf{R})$ is defined by an equation $F(x_0, \ldots, x_n) = 0$, where F is a homogeneous, degree-two polynomial in the homogeneous coordinates $(x_0, \ldots, x_n)$. Quadrics are classified by the rank and signature (p, q) of the form F, it being understood that quadrics with signatures (p, q) and (q, p) are projectively equivalent (change signs in the equation).

Borrowing the notations from section 2, the case where F has maximal rank $n + 1$ (that is, S has no multiple point) yields, when restricted to affine space:

(a) equation (62) with $r = n$ and $k \neq 0$, giving the homogeneous equation $a_1 x_1^2 + \cdots + a_n x_n^2 + k x_0^2 = 0$; or
(b) equation (63) with $r = n - 1$ and $b \neq 0$, giving the homogeneous equation $a_1 x_1^2 + \cdots + a_{n-1} x_{n-1}^2 + b x_n x_0 = 0$.

Case (b), which give parabolas and paraboloids, occurs when the hyperplane at infinity is tangent to S (always at the point $(0, 0, \ldots, 0, 1)$).

The discussion above holds over every field K of characteristic $\neq 2$. Since $4x_n x_0 = (x_n + x_0)^2 - (x_n - x_0)^2$, the equation in (b) reduces to that in (a) with a_n and k having opposite signs.

Let us then proceed to the classification of quadrics of maximal rank.

In the plane, the only inequivalent signatures are $(+++)$ and $(++-)$, the signs being those of a_1, a_2 and k, respectively. The first corresponds to the imaginary ellipse, the second to the three other curves: hyperbola, parabola and ellipse. Thus these three are projectively equivalent. What discriminates among them is the position of the line at infinity, which meets H at two real points, E at two complex conjugate points, and P at a single tangency point at infinity.

In three dimensions, the possible signatures are $(++++)$, $(+++-)$ and $(++--)$. The first gives the imaginary ellipsoid. The second gives the ellipsoid E, the two-sheet hyperboloid *H2* and the elliptic paraboloid EP. Their intersections with the plane at infinity are an imaginary ellipse, an irreducible real conic, and two complex conjugate lines of tangency, respectively.

The last signature gives the one-sheet hyperboloid *H1* and the hyperbolic paraboloid HP, which are precisely the quadrics which admit real lines (theorem 47). The existence of these lines is particularly easy to demonstrate if we write the equation of the quadric in the form $xt = yz$: one family, $\{D_p \mid p \in \hat{\mathbf{R}}\}$, is given by $x = py$, $z = pt$, and the other, $\{D_p^*\}$, by $x = pz$, $y = pt$.

Two generators D_p and D_q or D_p^* and D_q^* from the same family have no point in common; two generators D_p and D_q^* from different families have a unique common point, with homogeneous coordinates $(x, y, z, t) = (pq, p, q, 1)$. Every point of the quadric S is obtained in this way, and the lines D_p and D_q^* going through a point make up the intersection of S with its tangent plane at the point. Thus the map that takes $(p,q) \in \hat{\mathbf{R}} \times \hat{\mathbf{R}}$ to $D_p \cap D_q^*$ is a bijection $\hat{\mathbf{R}} \times \hat{\mathbf{R}} \to S$, showing that S can be seen as the product of two projective lines. We now generalize this result.

Segre varieties

Let $\mathbf{P}_n$ and $\mathbf{P}_q$ be n- and q-dimensional projective spaces over an arbitrary field K. We associate to the pair (X,Y), where $X \in \mathbf{P}_n$ and $Y \in \mathbf{P}_q$ have homogeneous coordinates $(x_0, \ldots, x_n)$ and $(y_0, \ldots, y_q)$, respectively, the point of $\mathbf{P}_{(n+1)(q+1)-1}$ with homogenous coordinates $z_{ij} = x_i y_j$, for $i = 0, \ldots, n$ and $j = 0, \ldots, q$. This point is well-determined by X and Y because multiplying all the x_i (or all the y_j) by the same factor simply multiplies the z_{ij} by this factor. The image of this correspondence lies in the subvariety S of $\mathbf{P}_{(n+1)(q+1)-1}$ defined by the $\frac{1}{4}n(n+1)q(q+1)$ equations

$$Z_{ij}Z_{i'j'} - Z_{ij'}Z_{i'j} = 0 \qquad \text{for every } i \neq i' \text{ and } j \neq j'. \tag{64}$$

Conversely, let the homogeneous coordinates in $\mathbf{P}_{(n+1)(q+1)-1}$ be indexed by (i,j), for $i = 0, \ldots, n$ and $j = 0, \ldots, q$, and consider a point (z_{ij}) of this space whose homogeneous coordinates satisfy (64). Some z_{ij} is non-zero; assume for concreteness $z_{00} \neq 0$. Let $X \in \mathbf{P}_n$ and $Y \in \mathbf{P}_q$ be the points with homogeneous coordinates $x_i = z_{i0}$ and $y_j = z_{0j}$, respectively; clearly X and Y do not depend on the choice of homogeneous coordinates for the point in $\mathbf{P}_{(n+1)(q+1)-1}$. Furthermore, if we choose another non-zero coordinate to play the role of z_{00} we arrive at the same X and Y: if $z_{uv} \neq 0$, we have $z_{i0}z_{0v} = z_{iv}z_{00}$ by (64), so (z_{i0}) and (z_{iv}) describe the same point in X (notice that both sets of homogeneous coordinates are non-zero because z_{00} and z_{uv} are), and similarly for (z_{0j}) and (z_{uj}). Finally, if we apply the previously defined correspondence $\mathbf{P}_n \times \mathbf{P}_q \to \mathbf{P}_{(n+1)(q+1)-1}$ to (X,Y), we get back the point we started with, since the products $x_i y_j = z_{i0}z_{0j}$ are equal to $z_{00}z_{ij}$ by (64).

This shows that the correspondence defined above is a bijection between $\mathbf{P}_n \times \mathbf{P}_q$ and the subvariety S of $\mathbf{P}_{(n+1)(q+1)-1}$ defined by equations (64). We call S the (n,q)*–Segre variety*, and the bijection the *Segre embedding*. We have embedded a product of projective spaces in a projective space.

In particular, if $n = q = 1$, we have four homogeneous coordinates z_{00}, z_{01}, z_{10} and z_{11}, so the Segre variety is embedded in $\mathbf{P}_3$. System (64) reduces to a single equation $z_{00}z_{11} - z_{01}z_{10} = 0$, so S is the quadric of equation $xt - yz = 0$ (returning to the more usual notation). The images in S of "horizontal" and "vertical" lines in $\mathbf{P}_1 \times \mathbf{P}_1$ are exactly the rectilinear generators D_p and D_p^* of S.

The product $\mathbf{P}_1 \times \mathbf{P}_2$ embeds in $\mathbf{P}_5$, and $\mathbf{P}_2 \times \mathbf{P}_2$ embeds in $\mathbf{P}_8$.

3.4. Classification of Conics and Quadrics over a Finite Field

We assume now that K is a finite field $\mathbf{F}_q$, and, for simplicity, that its characteristic is not 2. The next result is fundamental:

Theorem 48. *Every homogeneous, degree-two polynomial in at least three variables over* $\mathbf{F}_q$ *has a non-trivial zero. Except in dimension*0, *every quadric over* $\mathbf{F}_q$ *has a rational point.*

Proof. Writing the polynomial as a sum of squares of coordinates (see section 2) and making all but three of the coordinates vanish, we can restrict our attention to $ax^2 + by^2 + cz^2$. We can also assume $a, b, c \neq 0$, because if c, say, vanishes, $(0, 0, 1)$ is a solution.

Now let $z = 1$, so the equation to be solved becomes $by^2 + c = -ax^2$. Recall that $\mathbf{F}_q$ contains $(q-1)/2$ non-zero squares, or $(q+1)/2$ squares in all. As y runs over $\mathbf{F}_q$, the left-hand side of the equation takes on $(q+1)/2$ values, and so does the right-hand side as x runs over $\mathbf{F}_q$. Since $(q+1)/2 + (q+1)/2 > q$, the two sides have at least one value in common, which shows the existence of a solution $(x, y, 1)$. $\square$

This is a particular case of *Chevalley's theorem*: every homogeneous, degree-d polynomial in at least $d+1$ variables over a finite field has a non-trivial zero.

Classification of conics

If an irreducible conic has one rational point, it has others, by the parametric representation given in section 2.5. Here it has $q+1$ rational points in the projective plane. By choosing an appropriate projective frame (see theorem 16 in section 1.7, or the homogeneous parametric representation in section 2.5), the equation of the conic reduces to $x^2 - yz = 0$. Thus there is only one orbit of irreducible conics. In addition, since $\mathbf{F}_q$ has only one quadratic extension, there exist three types of reducible conics: two rational lines, two conjugate lines in the quadratic extension (with equation $x^2 - ny^2 = 0$, where n is a non-square in $\mathbf{F}_q$), and a double line.

In the affine plane, the classification of irreducible conics is based on the nature of the points at infinity: two rational points (*hyperbola*), two conjugate points in the quadratic extension (*ellipse*), or a double point, where the conic (a *parabola*) is tangent to the line at infinity. One can also work with equations (62) and (63) in section 2. For central conics, we normalize (62) to $ax^2 + by^2 - 1 = 0$. Bearing in mind that, in the decomposition of a quadratic form into squares, the first basis vector is entirely arbitrary (among non-isotropic vectors), we can take it to be a vector joining the center to a rational point of the curve; thus $ax^2 - 1 = 0$ has roots in $\mathbf{F}_q$, so a is a square and can be absorbed into x^2. Choosing a particular non-square n in $\mathbf{F}_q$, we can similarly normalize b to be 1 or n; the two possible equations are then $x^2 - y^2 - 1 = 0$ (hyperbola) and $x^2 - ny^2 - 1 = 0$ (ellipse). The parabola comes from equation (63): dividing $ax^2 + by = 0$ by a and applying an appropriate homothety to y we get $x^2 - y = 0$. Finally, there are six types of reducible conics: three split into concurrent lines and three into parallel lines.

Except for the inexistence of imaginary ellipses, the classification over $\mathbf{F}_q$ is the same as over $\mathbf{R}$.

Classification of quadrics in three dimensions

If such a quadric has a double point, we can project from that point and reduce the problem to the classification on the plane. Thus we will limit ourselves to quadrics without double points (which are, for that reason, irreducible).

Let S be such a quadric in projective space and A a rational point of S (theorem 48). With the exception of those that lie in the tangent plane T at A, the rational lines going through A intersect S again at another rational point (*stereographic projection*). Thus S has q^2 rational points in addition to T. Depending on the nature of the two lines that make up the intersection of T and S, then, we can have two cases:

(1) The two lines are rational, so their union has $2(q+1) - 1 = 2q + 1$ rational points. S has $q^2 + 2q + 1 = (q+1)^2$ rational points.

(2) They are quadratically conjugate, and their union has a single conjugate point A. In this case S has $q^2 + 1$ rational points.

Since $(q+1)^2 \neq q^2 + 1$, cases (1) and (2) cannot coexist on the same quadric. In case (1) there are, on S, two families of (rational) rectilinear generators, each containing $q + 1$ lines, each line containing $q + 1$ points. In this case S can be seen as the product of two projective lines (section 3); we say that S is a *ruled surface*.

To see that the two cases indeed exist and that all quadrics in each case are projectively equivalent, we set the origin at a rational point of S and take the tangent plane at the origin as the xy-plane. The homogeneous equation of S can then be written $P(x, y, z) + zt = 0$ with P homogeneous of degree two, or again $G(x, y) + z(ax + by + cz + t) = 0$, where G is homogeneous of degree two. After replacing t with $ax+by+cz+t$, we get the equation $G(x, y)+zt = 0$. By decomposing $G(x, y)$ into squares and fixing a non-square $n \in \mathbf{F}_q$, we obtain the two possible equations: (1) $x^2 - y^2 + zt = 0$, where the lines can be clearly seen if we write $(y - x)(y + x) = zt$; and (2) $x^2 - ny^2 + zt = 0$.

To summarize, there are two orbits of quadrics without double points in projective space: ruled quadrics and non-ruled quadrics.

The affine classification of quadrics without double points requires the study of the intersection of the quadric with the plane at infinity. Here this intersection, which is a conic defined over $\mathbf{F}_q$, always has a rational point (theorem 48). Thus it is either an irreducible conic or the union of two lines, in which case the plane at infinity is tangent to S, and S is called a *paraboloid*; otherwise S is a *hyperboloid*. Both paraboloids and hyperboloids can be ruled or non-ruled, for a total of four orbits.

One can recover this result algebraically, using equations (62) and (63) (section 2). For central quadrics, we normalize (62) to the form $ax^2 + by^2 + cz^2 - 1 = 0$. As for conics, theorem 48 allows us to assume that the first two axes of the decomposition intersect the quadric in rational points, that is, a and b are squares and can be absorbed into x^2 and y^2. The third axis is perpendicular to the first two, so it's uniquely determined, and the best we can do is normalize c to, say, -1 or $-n$, where $n \in \mathbf{F}_q$ is a fixed non-square. We get the following two equations:

- $x^2 + y^2 - z^2 - 1 = 0$ (ruled hyperboloid);
- $x^2 + y^2 - nz^2 - 1 = 0$ (non-ruled hyperboloid).

For paraboloids, equation (63) in the form $ax^2 + a'y^2 + bx = 0$ gives, after division by a, scaling along the z-axis and scaling along the y-axis, the following possibilities:

- $x^2 - y^2 - z = 0$ (ruled paraboloid);
- $x^2 - ny^2 - z = 0$ (non-ruled paraboloid).

The number of elements of these affine quadrics is:

- $(q+1)^2 - (q+1) = q^2 + q$ for the ruled hyperboloid,
- $(q^2+1) - (q+1) = q^2 - q$ for the non-ruled hyperboloid,
- $(q+1)^2 - (2q+1) = q^2$ for the ruled paraboloid,
- $(q^2+1) - 1 = q^2$ for the non-ruled paraboloid.

The number of points of the paraboloids can also be obtained from their equations $z = x^2 - y^2$ and $z = x^2 - ny^2$, because z is determined by x and y, which can take any value in $\mathbf{F}_q$.

CHAPTER 4

Polarity with Respect to a Quadric

The key ideas behind this whole chapter are very simple: orthogonality with respect to a symmetric bilinear form, and the isomorphism induced by a non-degenerate symmetric bilinear form over a vector space E between E and its dual.

We will assume that our field of scalars K has characteristic $\neq 2$ in order to skirt difficulties with the correspondence between quadratic forms and symmetric bilinear forms. Sometimes we will also need to assume K algebraically closed.

4.1. Polars and Poles

A quadric S in a projective space $\mathbf{P}(E)$ is defined by a homogeneous equation $Q(x) = 0$, where Q is a quadratic form. Let B be its associated bilinear form ($Q(x) = B(x, x)$). Two points $a, b \in \mathbf{P}(E)$ are said to be *conjugate* with respect to S if $B(a, b) = 0$ (throughout this chapter we will denote by the same letter a point of $\mathbf{P}(E)$, one of its representatives in E, and a set of homogeneous coordinates for it).

If B is non-degenerate, which means that S has no double point, each $x \in E$ can be associated a linear form $f(x)$ such that $f(x)(y) = B(x, y)$ for $y \in E$. The map f thus defined is an isomorphism between E and its dual E^*; it induces an isomorphism $\mathbf{P}(f)$ between $\mathbf{P}(E)$ and $\mathbf{P}(E^*)$, which is the space of hyperplanes of $\mathbf{P}(E)$ (section 1.5). The hyperplane $\mathbf{P}(f)(a)$ is called the *polar* of a; its equation is $B(a, x) = 0$.

If D is the vector line of E corresponding to a, the polar of a is the image of the orthogonal complement $D^{\perp}$ of D.

The isomorphism $\mathbf{P}(f)$ associates to every projective linear subspace $V' = \mathbf{P}(V)$ of $\mathbf{P}(E)$ a linear system of hyperplanes, whose base (theorem 9 in section 1.5) is $\mathbf{P}(V^{\perp})$; this projective linear space is said to be *conjugate* to V'. Conjugation with respect to S corresponds to orthogonality with respect to B. A projective linear space V' and its conjugate V'' satisfy the dimension formula

$$\dim(V') + \dim(V'') = \dim \mathbf{P}(E) - 1. \tag{65}$$

The conjugate of a hyperplane with respect to S is thus a point, called its *pole*. In dimension three, the conjugate of a line is a line. Conjugation, like orthogonality, is a symmetric relation.

Theorem 49. *Let S be a quadric without double points and a and a' points in $\mathbf{P}(E)$. Let m and m' be the points where $D_{aa'}$ intersects S. A necessary and sufficient condition for a and a' to be conjugate with respect to S is that either*

(a) *a and a' do not belong to S, and $(a, a', m, m') = -1$, that is, a and a' are harmonic conjugates with respect to m and m'; or*
(b) *$D_{aa'}$ is tangent to S at a (or a').*

Proof. The intersections m and m' correspond to the roots of the equation $Q(a + ta') = 0$, for $t \in \hat{K}$. This equation can be written $Q(a')t^2 + 2B(a, a')t + Q(a) = 0$. If $Q(a') \neq 0$ and $Q(a) \neq 0$, the condition $B(a, a') = 0$ means that the two roots add up to zero, that is, they are harmonic conjugates with respect to the parameters 0 and ∞ of a and a'; this is possibility (a). If $Q(a) = 0$, the equation has the root $t = 0$ and $B(a, a') = 0$ means that 0 is a double root, that is, $D_{aa'}$ is tangent to S. □

Corollary. *The polar of a point $a \in S$ is the tangent hyperplane to S at a.* □

Examples

(1) If the hyperplane at infinity is not tangent to an affine quadric S, its pole is the center of S. Otherwise S is called a *paraboloid*.

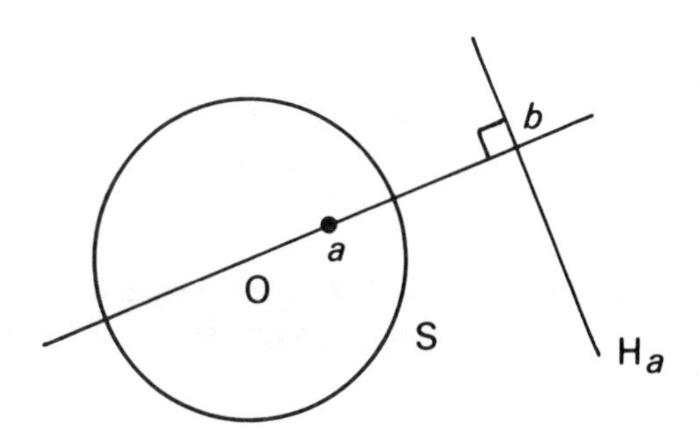

(2) The set of midpoints of chords of an affine quadric S parallel to a given direction is the polar of the point at infinity in that direction: it goes through the center of S or through the tangency point of S with the hyperplane at infinity.
(3) If S is a Euclidean sphere of center O and radius R, the polar of a point a is orthogonal to the line D_{Oa} and intersects that line at the point b such that $\overrightarrow{Ob}\overrightarrow{Oa} = R^2$.

4.2. Polarity with Respect to Conics

It is interesting to study the properties of polarity with respect to all the conics in a given pencil (section 1.7). Since some of these are degenerate, we must figure out what happens to the notions of the polar of a point and the pole of a line in these cases.

We should bear in mind that the relation $B(x, y) = 0$ is linear in each of the variables x, y, and also with respect to the coefficients of the form B, that is, with respect to the coefficients of the equation $Q(x) = 0$ of our conic. The notion of conjugate points always makes sense.

(1) For a conic that degenerates into two lines D and D', the relation $B(p, x) = 0$ is always satisfied when p is their intersection point. Otherwise $B(a, x) = 0$ is the equation of a line P_a going through p; the line is D (resp. D') when a is on D (resp. D'). When $a \neq D \cup D'$, theorem 49(a) still holds, so that the lines D, D', D_{pa} and P_a are in harmonic division (in the pencil of lines through a). To summarize, we can talk about the *polar* of a point $a \neq p$; this line always goes through p.

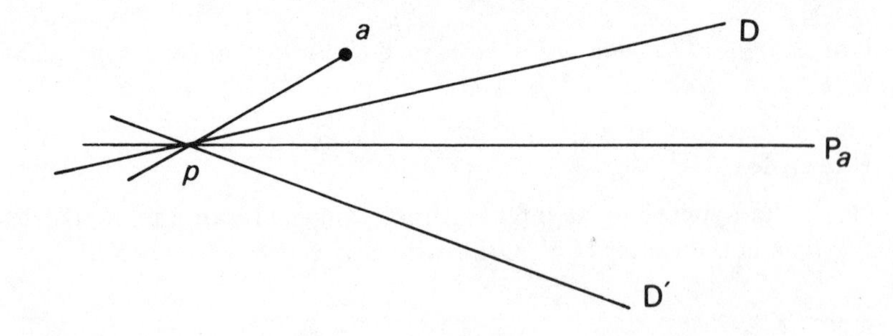

But a line E through p is the polar of all the points $a \neq p$ such that $(D, D', E, D_{pa}) = -1$, and a line that does not goes through p is the polar of no point at all. Thus the notion of the pole of a line does not make sense.

(2) When the conic is a double line $2D$, the equation $B(a, x) = 0$ is always satisfied if $a \in D$. If $a \notin D$, the equation $B(a, x) = 0$ is the equation of

D. Thus we can talk about the the polar of a point $a \notin D$, but talking about the pole of a line makes even less sense than in case (1).

We will denote by $P_C(a)$ the polar of a point a with respect to a conic C (of which a is not a double point). We propose now to study the polars of a point m with respect to the conics of a linear pencil F. If we let $Q(x) + tQ^0(x) = 0$ be the equation of a general conic $C(t) \in F$, the coefficients of the equation of $P_{C(t)}(m)$ are linear functions of t, so the polar either runs over a linear pencil of lines, or is fixed. Let's study the case when it is fixed, as a function of the type of the pencil F (see section 1.7).

(a) If F has four base points a, b, c, d (with no three on a straight line), it contains three degenerate conics $D_{ab} + D_{cd}$, $D_{ac} + D_{bd}$ and $D_{ad} + D_{bc}$. Denote by p, q, r their double points; we say that p, q, r is the *diagonal triangle* of the "quadrilateral" formed by a, b, c, d. If m is distinct from p, q, r, the polar P of m with respect to each of the three degenerate conics is well defined; if $P_{C(t)}(m)$ were fixed, it would have to pass through p, q, r, which are not on a straight line (because we assumed characteristic $\neq 2$).

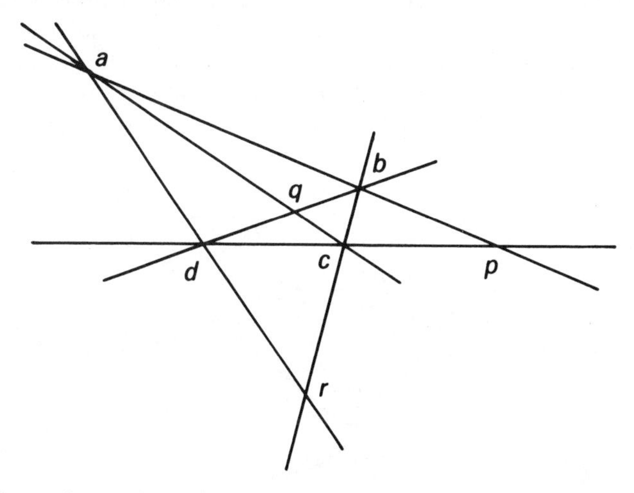

This can be verified by making a, b, c, d into a projective frame.

But, if $m = p$, the construction of the harmonic conjugate given in section 2.3 shows that D_{qr} is the polar of p with respect to the two degenerate conics that do not go through p. By linearity, D_{qr} is the polar of p with respect to all the conics in the pencil. The same is true for q and r. The three vertices of the diagonal triangle p, q, r are then pairwise conjugate with respect to each of the conics of F. A triangle whose vertices are pairwise conjugate with respect to a conic C is said to be *self-polar*; (p, q, r) is self-polar with respect to all the conics of F, and that it is the only such triangle.

In other words, if the conics corresponding to two quadratic forms $Q(x)$ and $Q^0(x)$ in three variables have four common points, there exists a basis orthogonal with respect to the two forms, and this basis is essentially unique.

The existence of a base orthonormal with respect to the inner product in $\mathbf{R}^3$ and Q-orthogonal can be deduced from this, if the conics $Q(x,y,z) = 0$ and $x^2 + y^2 + z^2 = 0$ have four distinct common points. For the points are pairwise conjugate, say $a, \bar{a}, b, \bar{b}$, and it is easy to see that each of the points $D_{a\bar{a}} \cap D_{b\bar{b}}$, $D_{ab} \cap D_{\bar{a}\bar{b}}$ and $D_{a\bar{b}} \cap D_{\bar{a}b}$ of the diagonal triangle is real. This basis is unique up to sign.

Remark. To construct a triangle p, q, r self-polar with respect to a conic C, one can take p arbitrary, q on the polar $P_C(p)$ and r the harmonic conjugate of q with respect to the points of $P_C(p) \cap C$.

(b) If F is made up of conics tangent to a line D at a and going through two other points b and c (type $(2,1,1)$), its degenerate conics are $D + D_{bc}$ and $D_{ab} + D_{ac}$; the double points are p and a. If a point m, distinct from a and p, has a fixed polar $P_{C(t)}(m)$, this can be no other than $D_{ap} = D$, and (1) above, applied to $D + D_{bc}$, forces m to be on D. But m cannot at the same time be on the harmonic conjugate of D with respect to D_{ab} and D_{ac}. Thus $m = a$ or $m = p$. The polar of a with respect to the true conics of pencil is their common tangent D, by theorem 49(b). That of p is the harmonic conjugate of D with respect to D_{ab} and D_{ac}. The pencil has no common self-polar triangle.

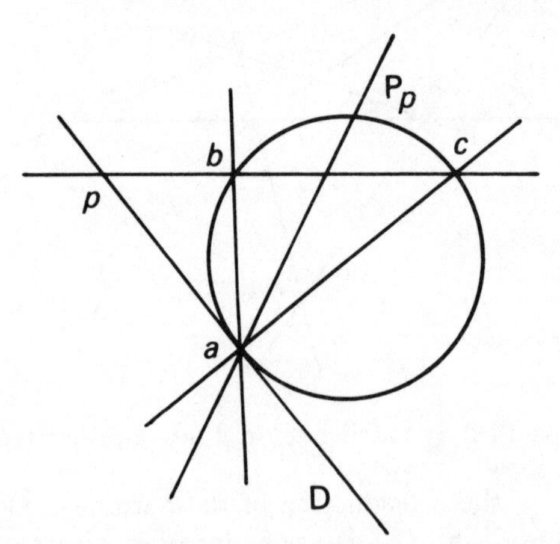

(c) If F is formed by the conics tangent to a line T at a and to a line U at b (type (2,2)), its degenerate conics are $T + U$ and $2D_{ab}$; its double points are p and all points of D_{ab}. If a point m outside D_{ab} has a fixed polar $P_{C(t)}(m)$, this polar can be no other than D_{ab}; thus $m = p$ because the pole of a line with respect to a proper conic C is the intersection of the tangents to C at the points where the line intersects C—a useful property stemming from theorem 49(b). If $m \in D_{ab}$, its polar with respect to the conics of the pencil is $D_{pm'}$, where m' is the harmonic conjugate of m with respect to a and b. The pencil has a whole family of common self-polar triangles, of the form (p, m, m') with m, m' on D_{ab} and $(a, b, m, m') = -1$.

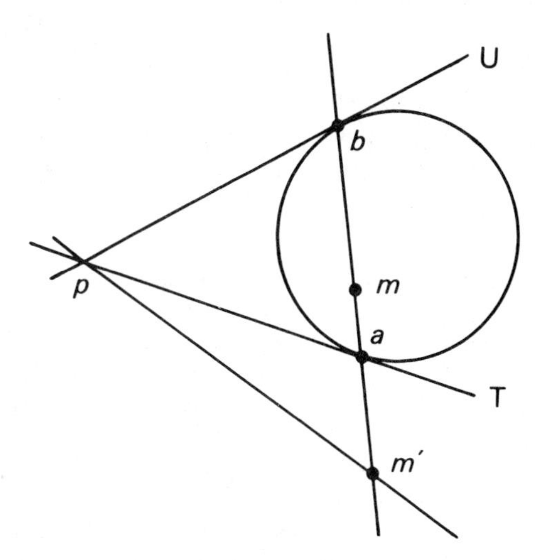

(d) If F is formed by conics osculating one another at a point a and going through another point b (type (3,1)), its only degenerate element is $D_{ab} + T$, where T is the common tangent to the conics at a; its double point is a. If a point $m \neq a$ has a fixed polar $P = P_{C(t)}(m)$, the polar goes through a, so m is conjugate to a and $m \in T$ (theorem 49(b)). Then the second intersection b' of D_{mb} with $C(t)$ is fixed, because it is the harmonic conjugate of b with respect to m and $m' = D_{mb} \cap P$; this is impossible. Thus a is the only point having a fixed polar, and that polar is clearly the common tangent T. The pencil has no common self-polar triangle.

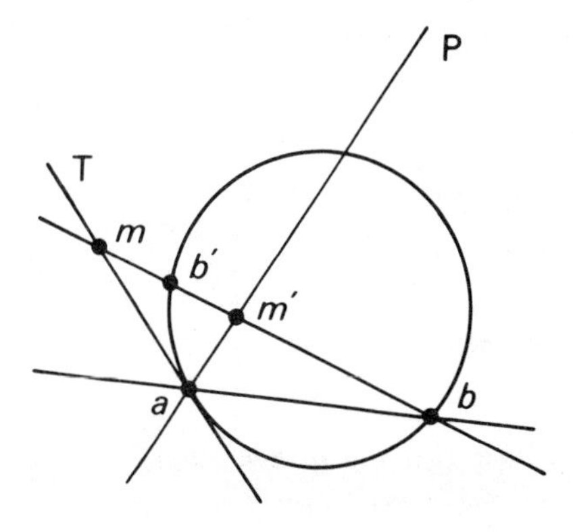

(e) If F is formed by the conics superosculating one another at a point a (type (4)), its only degenerate conic is $2T$, where T is the common tangent at a. If a point $m \notin T$ has a fixed polar, the polar can only be T; but, since the pole of T with respect to a proper conic $C(t)$ is a, we have $m = a$, which is impossible. On the other hand, a point $m \in T$ has a fixed polar, because, if we the equation of $C(t)$ in the form $xz + F(x,y) + tx^2 = 0$, where F is homogeneous of degree two (formula (38) in section 1.7), the polar of $m = (0, y_0, z_0)$ is $z_0 x + y_0 F'_y(x,y) = 0$, a line through $a = (0,0,1)$ which depends on m but not on t. If there is a common self-polar triangle, its three vertices have fixed polars, and so lie on T. By conjugation, they must be of the form m, a, a, which is impossible; thus there are not such triangles.

A word on pencils all of whose conics are degenerate:

(f) If they are of the form $D + D'$, where D and D' go through a fixed point a, D and D' are homologous under an involution j of the pencil of lines through a (theorem 31 in section 2.4). Denoting by F and F' the fixed lines of j, the points m having a fixed polar are those lying on F or F', and their polar is F' or F, respectively.

(g) If they are of the form $D + F$, where F is fixed and D goes through the pencil of lines that contain a point a, a point $m \notin F$ cannot have a fixed polar P: if $a \notin F$, such a polar would have to pass by the point $D \cap F$, which is variable, and if $a \in F$, D would be fixed, being the harmonic conjugate of D_{am} with respect to P and F. On the other hand, if $m \in F$, its polar is fixed.

We thus have the following theorem:

Theorem 50. *Let F be a pencil of conics containing proper conics. If m is a point in the plane, not equal to any of the double points of the degenerate conics of F, the map that takes each conic $C(t) \in F$ to the polar $P_{C(t)}(m)$ of m with respect to $C(t)$ is a projective transformation from F onto the pencil of lines going through a fixed point m'. The double points of the degenerate conics of F have fixed polars with respect to the conics of F. The conics admit one(resp. more than one) common self-polar triangles if and only if F has four distinct base points(resp. is a pencil of bitangent conics).* □

Remark. The conics to which a given triangle is self-polar form a dimension-two linear system: if we take the vertices of the triangle as the vertices of a projective frame, the equation of these conics is $ax^2 + by^2 + cz^2 = 0$. The conics with respect to which two points are conjugate (resp. a point and a line are pole and polar) form a four-dimensional (resp. three-dimensional) linear system.

Corollary. *The conics conjugate to a triangle and going through a fixed point p either pass through three other fixed points, or are bitangent at p*

and at some other point q. The second case occurs if and only if p is on one of the sides of the triangle.

Proof. By the remark, these conics form a pencil. Apply theorem 50 and look at the figures under items (a) and (c) above. □

Theorem 51. *The set of poles of a fixed line D with respect to the proper conics of a pencil F having four distinct base points is a conic S circumscribed to the common self-polar triangle (p, q, r).*

Proof. Take on D two points a, b distinct from p, q, r. As $C(t)$ ranges over F, the polars $P_{C(t)}(a)$ and $P_{C(t)}(b)$ range over pencils of lines, with base points a' and b', and they are in projective correspondence with each other (theorem 50). Thus their intersection, which is the pole of D with respect to $C(t)$, describes a conic S containing a' and b' (theorem 34 in section 2.5). If a is a point common to D and D_{pq}, its polars $D_{C(t)}(a)$ go through r, so that $a' = r$ and $r \in S$; the same holds for p and q. □

Strictly speaking, the set of these poles is S minus the three points that correspond to the values of t for which $C(t)$ is degenerate. These three points are p, q, r.

We leave to the reader the care of examining the case where S is degenerate (cf. theorem 34) and of extending theorem 51 to the other types of pencils of conics.

Corollary 1. *In the affine plane, the set of centers of conics circumscribed to a quadrilateral is a conic.*

Proof. Take for D the line at infinity. □

This conic has as points at infinity the points at infinity of the parabolas circumscribed to the quadrilateral. It goes through the midpoints of the six sides of the quadrilateral, for such a midpoint is conjugate, with respect to all the conics in the pencil, to the point at infinity on which it lies. Thus:

Corollary 2. *In the affine plane, the midpoints of the six sides of a quadrilateral and the three vertices of its diagonal triangle lie on the same conic.* □

When the quadrilateral is formed by the three vertices of a triangle plus the intersection of its altitudes, we saw in section 2.4 (an application of theorem 32) that the conics circumscribed to this quadrilateral are equilateral hyperbolas. The cyclic points are conjugate with respect to these hyperbolas. The conic in corollary 2 is then a circle; we recover the circle of nine points, which is also (corollary 1) the set of centers of equilateral hyperbolas circumscribed to the triangle.

Theorem 52. *Let (p,q,r) be a triangle self-polar with respect to a conic C and S a conic circumscribed to (p,q,r). Then every point $a \in S$ is the vertex of a triangle (a,b,c) inscribed in S and self-polar with respect to C.*

For readers not familiar with this language, a triangle (a,b,c) is inscribed in a curve S and S is circumscribed to (a,b,c) if the vertices a,b,c belong to S.

Proof. The polar $P = P_C(a)$ intersects S at two points b and c and C at two points m and m'. Let j and k be the intersections of P with D_{ap} and D_{qr}, respectively. Since D_{ap}, which joins the pole a of P and the pole p of D_{qr}, is the polar of k with respect to C, we have $(j,k,m,m') = -1$. Similarly, calling j', k', j'' and k'' the intersections of D with D_{aq}, D_{pr}, D_{ar} and D_{pq}, respectively, we have $(j',k',m,m') = (j'',k'',m,m') = -1$. Thus (j,k), (j',k') and (j'',k'') are pairs of homologous points of the involution g with fixed points m and m'. On the other hand, they are determined on P by intersection with the conics $D_{ap} + D_{qr}$, $D_{aq} + D_{pr}$ and $D_{ar} + D_{pq}$, which belong to the pencil F with base points a,p,q,r. But the conics of F determine an involution on P by the Desargues-Sturm theorem (theorem 31, section 2.4) which, having three pairs in common with g, must be g. Since the conic S is in F and determines the pair (b,c) on P, we see that $(c,b,m,m') = -1$. In other words, b and c are conjugate with respect to C and (a,b,c) is self-polar with respect to C. □

We say sometimes that S is *harmonically circumscribed* to C.

Corollary 1. *The vertices of two triangles self-polar with respect to a conic C lie on the same conic. Conversely, if a,b,c,a',b',c' are six points on the same conic, the triangles (a,b,c) and (a',b',c') are self-polar with respect to some conic C.*

Proof. If (a,b,c) and (a',b',c') are self-polar with respect to C, consider the conic S going through a,b,c,a',b': the polar $P_C(a)$ intersects S at two points, one of which is b and the other, c_1, is the pole of D_{ab} with respect to C, by theorem 52. But this is c, by assumption; thus $c = c_1$ is on S.

Conversely, the conics conjugate to (a',b',c') form a two-dimensional linear system. Requiring that a and D_{bc} be pole and polar with respect to it adds two linear conditions, so at least one conic C answers all these conditions. By theorem 52, the triangle (a,b,c) is conjugate to C. □

Corollary 2. *In the Euclidean plane, a circle harmonically circumscribed to an equilateral hyperbola H goes through its center. The set of centers of equilateral hyperbolas with respect to which a triangle is self-polar is the circle circumscribed to this triangle.*

Proof. Since the cyclic points i and j are conjugate with respect to H, they are the vertices of a triangle (i,j,c) self-polar with respect to H, and c is the center of H, being the pole of the line at infinity D_{ij}. This gives the first assertion, by corollary 1; the second is a consequence of the first. □

4.3. Polarity and Tangential Equations

As in section 1, we consider a quadric S without double point in a projective space $\mathbf{P}(E)$, and its homogeneous equation $Q(x) = 0$, where Q is a non-degenerate quadratic form. The bilinear form B associated with Q defines an isomorphism $f : E \to E^*$, given by $f(x)(y) = B(x, y)$. We can take the push-forward of Q under f, that is, the quadratic form Q^0 on E^* such that

$$Q^0(u) = Q\big(f^{-1}(u)\big) \qquad \text{for every } u \in E^*; \tag{66}$$

we call Q^0 the *inverse form* of Q. Its associated bilinear form B^0 satisfies $B^0(u, v) = B\big(f^{-1}(u), f^{-1}(v)\big)$.

Theorem 53. *A hyperplane H with equation $u(x) = 0$ is tangent to S if and only if $Q^0(u) = 0$.*

Proof. The hyperplane tangent to the point $x_0 \in S$ has equation $B(x_0, x) = 0$, that is, $f(x_0)(x) = 0$. If we write it in the form $u(x) = 0$, this condition means that $u \in f(S)$, $f^{-1}(u) \in S$, $Q\big(f^{-1}(u)\big) = 0$ and finally $Q^0(u) = 0$. □

The equation $Q^0(u) = 0$ is called the *tangential equation* of S.

Corollary. *If $M = \big(B(e_i, e_j)\big)$ is the matrix of B in some basis (e_i) of E, the matrix of the inverse form in the dual basis of E^* is M^{-1}.*

Proof. For $x, y \in E$, write $B^0\big(f(x), f(y)\big) = B(x, y)$; denote by M^0 the matrix of B^0 in the dual basis and by X and Y the column vectors corresponding to x and y. Then we have ${}^tX\,{}^tM(f)M^0M(f)Y = {}^tXMY$, where $M(f)$ is the matrix of f with respect to the given basis of E and the dual basis of E^*. Since $f(e_i)(e_j) = B(e_i, e_j)$, the matrix $M(f)$ is M, and we have ${}^tM = M$ because B is symmetric. Thus ${}^tXMM^0MY = {}^tXMY$ for all column vectors X and Y, which gives $MM^0M = M$, whence $M^0 = M^{-1}$ because M is invertible. □

This corollary is useful in computing the tangential equation of a quadric. The relation $Q^0(u_0e_0 + \cdots + u_ne_n) = 0$ expresses the fact that the hyperplane $u_0x_0 + \cdots + u_nx_n$ is tangent to S.

Examples

(1) The tangential equation of the conic $a_0x_0^2 + \cdots + a_nx_n^2 = 0$ is

$$a_0^{-1}u_0^2 + \cdots + a_n^{-1}u_n^2 = 0.$$

(2) In the plane, let S be the conic $ayz + bzx + cxy = 0$. We have

$$M = \begin{pmatrix} 0 & a & b \\ c & 0 & a \\ b & a & 0 \end{pmatrix}$$

and $\det M = 2abc$; computing the inverse of M gives

$$au^2 + bv^2 + cw^2 - abuv - acuw - bcvw$$

for the tangential equation of S.

(3) Still in the plane, it can be more convenient to express the fact that a line $ux + vy + wz = 0$ is tangent to a conic $F(x, y, z) = 0$ by writing that the homogeneous equation $F(x, y, -(ux + vy)/w) = 0$ in x and y, or, better yet, $F(wx, wy, -ux - vy) = 0$, has a double root, which is expressed by the discriminant being 0.

This last procedure can be generalized. We will limit ourselves to the plane for simplicity. Let C be an algebraic curve in the projective plane, with homogoneous equation $F(x, y, z) = 0$. As above, the condition that a line $ux + vy + wz = 0$ be tangent to C (or pass through one of its multiple points) is equivalent to the vanishing of a discriminant, so it an algebraic condition and homogeneous in u, v, w. By getting rid of factors $au + bv + cw$ which express the passing through a multiple point (a, b, c), we obtain a homogeneous polynomial equation $F^0(u, v, w) = 0$ which we call the *tangential equation* of C. Its degree is called the *class* of C.

When F has degree n, the study of the discriminant of a degree-n equation shows that, after getting rid of the linear factors corresponding to the multiple points of C, the degree of the relation between u, v and w is $n(n-1)$. If C has, as multiple points, exactly d nodes and r ordinary cusps, one can show that the class of C is $c = n(n-1) - 2d - 3r$ (*Plücker's formula*).

Thus a cubic without multiple points has class six, a nodal cubic class four and a cuspidal cubic class three (cf. section 2.6).

The class of a plane curve C is thus, by duality (section 1.5), the number of tangents (real or not, distinct or not) that can be drawn to C from an arbitrary point in the plane.

By theorem 53, a conic without double points is a curve of second class.

The set E of lines $ux+vy+wz = 0$ in the projective plane whose "coordinates" (u, v, w) satisfy a homogeneous polynomial equation $G(u, v, w) = 0$ of degree c is sometimes called an *envelope of class c*. An envelope of second class is also called a *tangential conic*. If G has no linear factors, every point of the plane lies on c lines (real or not, distinct or not) belonging to this envelope.

Applying to G the discriminant calculation that allowed us to pass from the punctual equation $F(x, y, z) = 0$ of a curve C to its tangential equation $F^0(u, v, w) = 0$, we obtain a homogeneous equation $G^0(x, y, z) = 0$. This equation expresses the fact that, among the lines of the envelope E that go through the point (x, y, z), at least two coincide.

When $G(u, v, w) = 0$ is the tangential equation $F^0(u, v, w) = 0$ of a curve C whose punctual equation is $F(x, y, z) = 0$, it is not at all clear that $G^0(x, y, z) = 0$ is the punctual equation $F(x, y, z) = 0$ we started with, that is, that $F^{00} = F$ up to a multiplicative factor. This is in fact false in characteristic $p \neq 0$ because of curves like $y = x^p$, all of whose tangents go through the same point.

This fact is, however, true, in characteristic $\neq 2$, when C is a conic without double points, because the inverse form of an inverse form is the original form $F^{00} = F$ (cf. theorem 53).

Let us state the general problem more precisely. Recall from section 1.3 that the tangent T at a simple point $P = (a, b, c)$ of a curve C of equation $F(x, y, z) = 0$ is the line through P that intersects C with multiplicity ≥ 2 at P; its equation is $xF'_x(a, b, c) + yF'_y(a, b, c) + zF'_z(a, b, c) = 0$ (formula (23)). Rephrasing that for an envelope E with equation $G(u, v, w) = 0$, we take a line D "simple" in E (that is, one that does not make the three partial derivatives G'_u, G'_v and G'_w vanish), and we look for the point K of D for which, among the lines of E going through K, several coincide at D; its homogeneous coordinates are $\big(G'_u(p, q, r), G'_v(p, q, r), G'_w(p, q, r)\big)$, if $px + qy + rz = 0$ denotes the equation of D. This is called that the *characteristic point* of D. The precise problem is then the following: if E is the set of tangents to a curve C and D is the tangent to C at a point P, is the characteristic point K of D the contact point P?

When the field of scalars K is locally compact, in particular $\mathbf{R}$ or $\mathbf{C}$, and the projective plane is given the compact topology described in section 2.8, just as the tangent at P to a curve C is the limit of the line joining P to a neighboring point P', so is the characteristic point of the tangent D the limit of the intersection I of D with the tangent D' at a neighboring point.

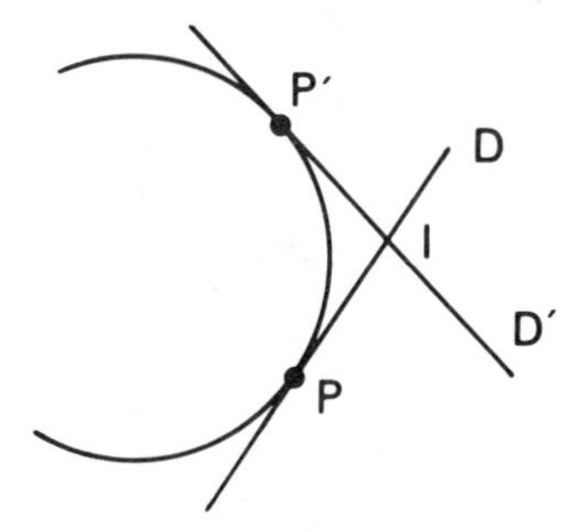

If we take P as the origin of affine coordinates and the tangent D as the x-axis, C can be described near the origin by a series expansion $y = ax^k + bx^{k+1} + \cdots$, where $a \neq 0$. The equation of the tangent D' at the point (x, y) of C is $Y - y = y'(X - x)$; it intersects D, the line $Y = 0$, at the point with abscissa $X = x - y/y'$. If the exponent k of the lowest-degree term of y is not a multiple of the characteristic (always the case in characteristic zero), y/y' is approximately x/k when x approaches 0, and the x-coordinate of I tends to 0. Thus we have the following result:

Theorem 54. *When the field of scalars is locally compact and has characteristic zero(that is,* $\mathbf{R}$, $\mathbf{C}$ *or a finite extension of* $\mathbf{Q}_p$*), the characteristic point of the tangent to an algebraic curve is its contact point. In particular, if we denote by* $F^0(u,v,w)$ *the tangential equation of a curve whose punctual equation if* $F(x,y,z) = 0$*, the equations* $F^{00}(x,y,z) = 0$ *and* $F(x,y,z) = 0$ *are equivalent.* □

One can even show that the two equations are proportional (if F is irreducible, this just means that F^{00} is not a power of F; see section 1.8). This will be the case if one takes the precaution of "cleaning up" F^0, by removing multiple or parasite factors.

Since each of the assertions in theorem 54 involves only a finite number of elements in the field of scalars, and since every extension of $\mathbf{Q}$ of finite type can be embedded in $\mathbf{C}$, theorem 54 holds for every field of characteristic zero.

The remarks above are useful in the study of polarity with respect to a conic. We are given a conic without double point S in the projective plane, with equation $Q(x,y,z) = 0$. To every point a of the plane, we associate its polar $P(a)$ with respect to S, and to every line D its pole $p(D)$; by section 1, these two maps are mutually inverse isomorphisms between the projective plane and its dual. Collinear points are taken into concurrent lines, and vice versa.

There is no harm in general, and it is convenient for the Euclidean properties, to assume that S has equation $x^2 + y^2 - z^2 = 0$. Then the point with coordinates (x,y,z) and the line with coefficients (u,v,w) correspond under this transformation if and only if we have, up to a factor:

$$x = u, \qquad y = v, \qquad z = -w. \tag{66}$$

(If S has a more general equation, u, v and w are linear forms, of x, y and z, and conversely: for instance, $u = Q'_x(x,y,z)$.)

Let C be an algebraic curve of degree d and class c, having $F(x,y,z) = 0$ and $F^0(u,v,w) = 0$ as its punctual and tangential equations, respectively. We associate to C:

- on the one hand, the envelope E' formed by the polars of points of C; its equation is $F(u,v,-w) = 0$ by (66), and its class is d;
- on the other hand, the curve C' formed by the poles of tangents to C; its equation is $F^0(x,y,-z) = 0$, and its degree is c.

In characteristic 0 (also in characteristic $\neq 2$ if C is a conic), E' is the set of tangents to C' (and, dually, C' is the set of characteristic points of E') by theorem 54: indeed, the tangential equation of C' is $F^{00}(u,v,-w) =$

0, which is equivalent to $F(u,v,-w) = 0$. We say that C' is the *polar transform* of C (with respect to S). We see right away that C is also the polar transform of C'. Class and degree are switched.

Examples. If C is a cuspidal cubic ($d = 3$, $c = 3$), so is C'. If C is a nodal cubic ($d = 3$, $c = 4$), C' is a quartic of third class, which, by Plücker's formula, is a quartic with three cusps ($3 = 4(4-1) - 3 \cdot 3$).

If C is unicursal, so is C', because from the homogeneous parametric representation $\big(P(t), Q(t), R(t)\big)$ of C, we deduce that the tangent to C at the point with parameter t has equation

$$\big(Q(t)R'(t)-R(t)Q'(t)\big)x+\big(P(t)R'(t)-R(t)P'(t)\big)y+\big(P(t)Q'(t)-Q(t)P'(t)\big)z=0.$$

Thus $(QR' - RQ', RP' - PR', -PQ' + QP')$ is a homogeneous parametric representation. It seems to be of degree $d(d-1)$, where $d = \max(d^0P, d^0Q, d^0R)$ (which equals d^0C if the representation is proper), but there are simplifications.

Thus, for the nodal cubic (t, t^2, t^3+a), we obtain for C' the representation $(t^4 - 2at, -2t^3 + a, t^2)$, which is indeed a quartic. If C is the cuspidal cubic $(1, t, t^3)$, we obtain $(2t^3, -3t^2, -1)$, another cuspidal cubic (set $t = 1/u$).

In the Euclidean plane, we saw in section 1 that, with respect to the unit circle $x^2 + y^2 - 1 = 0$, the pole of a line D is the inverse a, under the inversion with pole at the origin and power 1, of the orthogonal projection b of the origin on D. As D ranges over the set of tangents to a curve C, the curve described by the orthogonal projection b of a fixed point P on D is called the *pedal curve* of P with respect to C; for example, the pedal of the focus of a parabola with respect to the parabola is the tangent at the vertex. Thus the polar transform C' is the inverse of the pedal of the origin with respect to C.

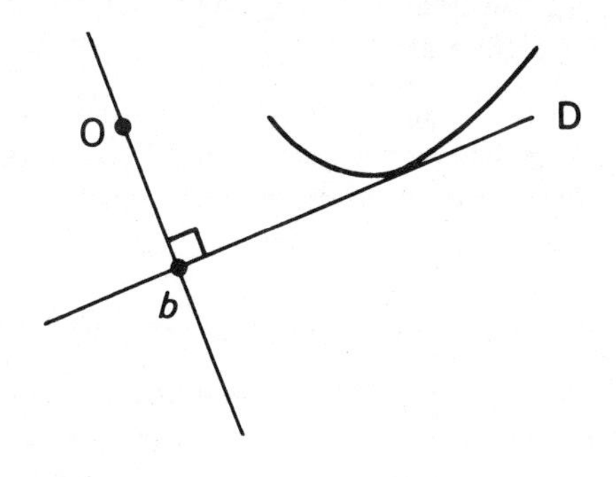

4.4. Applications to Conics

We start with immediate applications of theorem 53, which says essentially that a proper conic is an envelope of second class, to punctual results on conics proved in section 2.5. The new results can be thought of as the translations of the old ones into the dual language of tangents:

(a) Translation of theorem 34 about the generation of conics by two pencils between which there is a projective transformation: Let h be a projective transformation from a line D onto a line D'. As m runs over D, the line $D_{mh(m)}$ envelopes a conic C tangent to D and D', as long as the point a common to D and D' is not fixed under h. If it is fixed, the lines $D_{mh(m)}$, for $m \neq a$, all go through a fixed point.

If, in the affine plane, h is affine, the point at infinity of D' is the image under h of the point at infinity of D, and the envelope C is a parabola.

(b) Translation of Frégier's theorem (theorem 35): Let j be an involution on a proper conic C. As m ranges over C, the intersection point of the tangents to C at m and $j(m)$ describes a line D, which goes through the fixed points of j.

In the affine plane, a parabola P admits a single tangent with direction D, and this tangent is a projective function of D, because the other tangent to P drawn from the point at infinity of D is the line at infinity, and is fixed. Thus, in the Euclidean plane, the points of P where the tangents are orthogonal are in involutive correspondence. Thus the intersection of these tangents describes a line, which goes through the points of P where the tangents are isotropic; we will see that this line is the directrix of P.

(c) Translation of Pascal's theorem (theorem 35): Let T_i, for $i = 1, \ldots, 6$, be tangents to a proper conic C. If we call i the intersection point $T_i + T_{i+1 \ (\mathrm{mod}\ 6)}$, the lines D_{14}, D_{25} and D_{36} are concurrent. This is called *Brianchon's theorem.*

In particular, when a conic C is tangent to the three sides of a triangle (a, b, c), the lines joining the vertices a, b, c to the contact points of the opposite sides are concurrent (cf. corollary to theorem 36).

Correspondence between punctual and tangential decompositions

In view of results involving degenerate punctual conics, we study the degeneration of tangential conics. In the same way that a puncual conic can degenerate into two lines $D + D'$, or into a double line $2D$, a tangential conic, with equation $G(u, v, w) = 0$, can degenerate into either:

- two pencils of lines (G is the product of two linear forms), in which case we say that it reduces to the two base points b and b', and denote it by $b + b'$; or

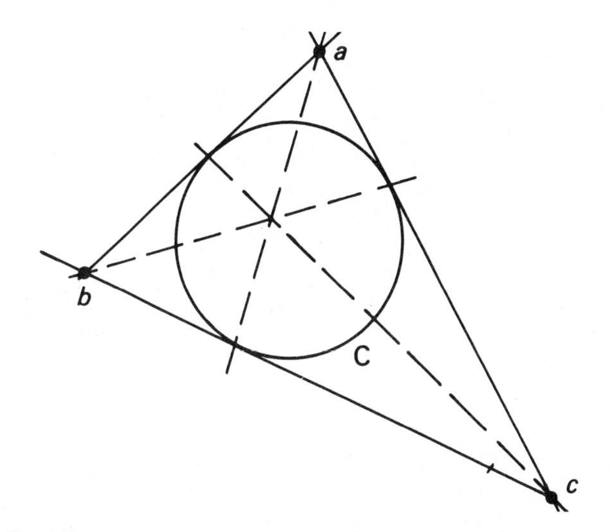

- one pencil of lines, counted twice (G is the square of a linear form), in which case we say that it reduces to a double point b, the base point of the pencil, and denote it by $2b$.

Theorem 53 gives a (bijective) correspondence between non-degenerate punctual conics and non-degenerate tangential conics. We will extend this correspondence to degenerate conics. Denote by P and P^0 the five-dimensional projective spaces of punctual and tangential conics, respectively. We will define the graph in $P \times P^0$ of the desired correspondence. Write the coefficients of the punctual conic

$$ax^2 + by^2 + cz^2 + 2a'yz + 2b'zx + 2c'xy = 0$$

as the matrix $M = \begin{pmatrix} a & c' & b' \\ c' & b & a' \\ b' & a' & c \end{pmatrix}$ of the associated bilinear form. Similarly, $M^0 = \begin{pmatrix} A & C' & B' \\ C' & B & A' \\ B' & A' & C \end{pmatrix}$ will denote the coefficients of a tangential conic

$$Au^2 + Bv^2 + Cw^2 + 2A'vw + 2B'wu + 2C'uv = 0$$

By theorem 53, this graph is defined, for proper conics, by the relation

$$MM^0 \text{ is proportional to } I, \tag{67}$$

where I is the identity matrix.

The relation $MM^0 = I$ is not homogeneous and we would not be able to extend it.

More explicitly, this relation is expressed by six equations of the type

$$aC' + c'B + b'A' = 0 \tag{68}$$

for the off-diagonal terms, and by

$$(69) \qquad aA + c'C' + b'B' = c'C' + bB + a'A' = b'B' + a'A' + cC$$

for the diagonal terms.

One may notice that these equations become linear if we identify $P \times P^0$ with the Segre variety in $\mathbf{P}_{35}$ (section 3.3).

For a conic made up of two lines, $D + D'$, which we can assume to be $xy = 0$, only c' is non-zero. Equations (68) then give $B = 0$, $A' = 0$, $A = 0$, $B' = 0$, and equations (69) give $C' = C' = 0$. Thus only C is non-zero in M^0, and a unique tangential conic, $w^2 = 0$, corresponds to $D + D'$; this conic is twice the intersection point of D and D'.

For a double line $2D$, which we can assume to be $x^2 = 0$, only a is non-zero. Equations (68) reduce to $C' = B' = 0$ and (69) to $A = 0 = 0$. The tangential conics corresponding to $2D$ have equation $Bv^2 + Cw^2 + 2A'vw = 0$. They amount to pairs of points lying on D.

Thus the extended correspondence is defined by:

- to two lines $D + D'$ we associate the intersection $D \cap D'$, counted twice;
- to a double line $2D$ we associate all pairs of points on D.

By symmetry,

- to a pair of points $b + b'$ we associate $2D_{bb'}$; counted twice;
- to a double point $2b$ we associate all conics $D + D'$, where D and D' are lines containing b.

Tangential pencils

A tangential pencil is the set of tangential conics $E(t)$ whose equation is of the form $F(u, v, w) + tG(u, v, w)$, for $t \in \hat{K}$. Such a pencil admits a finite number of *base lines*, tangent to all the conics in the pencil. Any line distinct from the base lines is tangent to a unique conic in the pencil (cf. section 1.7).

The most general example of a tangential pencil is the set of conics tangent to four lines, no three of which are collinear (cf. section 1.7, theorem 18). The other types of tangential pencils (containing proper conics) are defined by four conditions: some specifying contact with given lines and other passage through given points. For example, we can specify that the conics be tangent to a given line at a given point, and tangent to two other lines (type $(2, 1, 1)$). Notice that conditions that involve points and line symmetrically define tangential pencils which are at the same time punctual pencils. There are two such types of pencils:

- type $(2, 2)$: conics tangent to two given lines D, D' at two given points a, a';
- type (4): superosculating conics, tangent to a given line at a given point.

Even in these two cases, of course, the degenerate conics of the punctual and tangential pencils are not the same. In type $(2, 2)$ the degenerate punctual conics are $D + D'$ and $2D_{aa'}$, and the degenerate tangential conics are $a + a'$ and $2(D \cap D')$.

In the tangential pencil of conics tangent to four given lines, there are three degenerate conics, formed by the pairs of "opposite vertices" of the "quadrilateral" formed by the four lines. More formally, there are six intersection points of pairs of the four lines A, B, C, D; the degenerate conics are unions of two points from complementary pairs: for example, $(A \cap C) + (B \cap D)$.

We see also (cf. section 1.7, theorem 17) that there exists a unique tangential conic tangent to five distinct lines (no four of which are concurrent). This conic is proper if and only if no three of the lines are concurrent.

With respect to a proper conic S, the notions of conjugate points, conjugate lines, pole and polars, understood in the punctual and tangential sense, coincide: in fact (cf. beginning of section 3), orthogonality with respect to the quadratic form Q is mapped to orthogonality with respect to the inverse form Q^0. This is, in fact, the key idea in the study of polarity.

This allows us to translate the main results in section 2. The notion of the pole of a line with respect to a (possibly degenerate) tangential conic always makes sense, but the notion of the polar doesn't always. We have:

(1) Translation of theorem 50: If a line D is not one of those that join the points of the degenerate conics in a tangential pencil F, the poles of D with respect to the conics of F describe a line D', and the map thus defined from F onto D' is a projective transformation.

In particular, in the affine plane, we can take as D the line at infinity and conclude that the centers of conics tangent to four given lines describe a line D'. This line goes through the three midpoints of "opposite sides" of the quadrilateral formed by the four lines; in particular, the midpoints are collinear.

(2) Translation of theorem 51: Let F be the tangential pencil of conics tangent to the four sides of a (true) quadrilateral. The polars of a fixed point p with respect to the proper conics of F are tangent to a conic C, which is also tangent to the three diagonals (the lines joining opposite vertices) of the quadrilateral.

In particular, in the affine plane, when the quadrilateral is a parallelogram (two pairs of parallel lines), C is a parabola tangent to the diagonals of the parallelogram.

Foci

In the Euclidean plane, a *focus* of an algebraic curve C is any point f such that the two isotropic lines through f are tangent to C.

A curve of class c has, in general, c^2 foci, because we can draw c tangents to C through each of the cyclic points. Thus a proper conic has, in general, four foci. This number decreases if the line at infinity (which joins the cyclic points) is tangent to C (in which case there are $(c-1)^2$ foci), or if C goes through the cyclic points (if they are simple points of the curve, the tangents to C there count twice, so there are $(c-1)^2$ foci).

The decrease in the number of foci is greater if the line at infinity has a higher-order contact with C, or if the cyclic points are multiple points of C.

In particular, a parabola has a single focus; a circle has a single focus, its center (the asymptotes of a circle are the isotropic lines through its center).

Let us study the four foci of a central conic C (with real coefficients) that is not a circle. Let T and U be the tangents to C drawn from the cyclic point i; the tangents drawn from the other cyclic point $\bar{i}$ are the conjugate lines $\bar{T}$ and $\bar{U}$. The intersection points $T \cap \bar{T}$ and $U \cap \bar{U}$ are two real foci; the other two are complex conjugate foci. Reflection through the axes of the conic exchanges the cyclic points, the lines T and $\bar{T}$, and U and $\bar{U}$; it follows that the real foci lie on one of the axes of C and the imaginary foci on the other. Both pairs are symmetric with respect to the center.

Similarly, the focus of a parabola lies on its axis.

The two real (or imaginary) foci determine the other two, as the non-cyclic intersection points of the isotropic lines going through the given foci.

The expression of the coordinates (x, y) of the foci of a conic C given by its equation is particularly easy if the equation is tangential, $G(u, v, w) = 0$. For the isotropic line $Y - y = i(X - x)$ has coordinates $u = -i$, $v = 1$, $w = ix - y$, whence the equation $G(-i, 1, ix - y) = 0$; the other isotropic line gives the equation $G(i, 1, -ix - y) = 0$. Naturally, if G has real coefficients, we can replace these equations by their sum and difference.

For example, the ellipse $x^2/a^2 + y^2/b^2 - 1 = 0$ has tangential equation $a^2u^2 + b^2v^2 - w^2 = 0$ (see beginning of section 3), whence the equations $-a^2 + b^2 - (ix + y)^2 = 0$ and $-a^2 + b^2 - (ix - y)^2 = 0$. These are equivalent to $4ixy = 0$ and $2(-a^2 + b^2 + x^2 - y^2) = 0$. If the x-axis is the major axis (that is, $a > b$) and we set $c = \sqrt{a^2 - b^2}$, we obtain the four foci $(c, 0)$, $(-c, 0)$, $(0, ic)$ and $(0, -ic)$. Thus the real foci are the well-known foci along the major axis.

For the hyperbola $x^2/a^2 - y^2/b^2 - 1 = 0$ the tangential equation is $a^2u^2 - b^2v^2 - w^2 = 0$. Setting $c = \sqrt{a^2 + b^2}$, we obtain four foci with the same coordinates $(c, 0)$, $(-c, 0)$, $(0, ic)$ and $(0, -ic)$. The real foci lie on transverse axis and are well-known.

The parabola $x^2 - 2py = 0$ has tangential equation $pu^2 - 2vw = 0$. The equations of the foci are $-p - 2(ix - y) = -p + 2(ix + y) = 0$, and admit only one solution $(0, \frac{1}{2}p)$, which is well-known.

Conics sharing two given foci f and f', where f and f' do not lie along the same isotropic line (this is the case if they are real points) are said to be *homofocal*. They form a one-dimensional tangential pencil H, being the conics tangent to the four isotropic lines of f and f'. The degenerate conics in H are $f+f'$, $i+\bar{i}$ (the cyclic points) and $g+g'$ (the other two foci). The conics of H all have the same center, the midpoint of ff'.

Let m be a point in the plane and F_m the pencil of lines going through m. The tangential translation of the Desargues–Sturm theorem (theorem 31 in section 2.4) shows that the two tangents drawn from m to a variable conic of H are under involutive correspondence $j : F_m \to F_m$. Thus there are two conics of H going through m; those whose tangents are m are the fixed lines T and U of j. If f and f' are real, one of these conics is an ellipse and the other a hyperbola. Because of the degenerate conic $i+\bar{i}$, the isotropic lines of m are homologous under j, so T and U are orthogonal (section 2.4). Since D_{mf} and $D_{mf'}$ are also homologous, they are symmetric with respect to T and U, which are thus the bisectors of the lines D_{mf} and Dmf'. We have recovered some well-known properties.

The polar with respect to a conic C of a focus f of C is called the *directrix associated with this focus*. It joins the points a, a' where the isotropics I, I' through f are tangent to C. Thus C is part of the punctual pencil of conics tangent to I at a and to I' at a'. This pencil contains $I+I'$ and $2D_{aa'}$. Making f the origin and letting $px+qy+r=0$ be the equation of the directrix $D_{aa'}$, the equation of C is of the form

$$x^2+y^2-t(px+qy+r)^2=0. \tag{70}$$

This equation expresses the fact that C is the set of points whose distances to the focus f and to the associated directrix D are in a constant ratio.

This ratio is $\sqrt{t(p^2+q^2)}$. It is easy to see that this equals 1 if and only if the sum of quadratic terms in (70) is a square, that is, if and only if C is a parabola.

Notice that the distance $d(f,m)$ from the focus f to a variable point on the conic C is, up to a sign, an affine function of the coordinates (x,y) of m. More generally, suppose we have a real curve C and a point f such that the distance $d(f,m)$ from f to a variable point $m \in C$ is a rational function $R(x,y)$ of the coordinates (x,y) of m (up to sign). If f is the origin and R is the quotient P/Q of two polynomials, we have

$$(x^2+y^2)Q(x,y)^2-P(x,y)^2=0. \tag{71}$$

This is not an identity that holds for every x and y, because x^2+y^2 is not a square. If one chooses R (which is only determined modulo the equation of C), then P and Q, in such a way that $\max(d^0P, d^0Q)$ is minimal, one can show that the left-hand side of (71) is irreducible, so it is the equation of C. This shows that the intersections of C with the isotropic lines $y=ix$ and $y=-ix$ of f all have even multiplicity. But not all the foci of plane algebraic curves have this property. Equation (71) easily shows that such a curve must have even degree (which excludes cubics, even circular ones, even nodal or cuspidal ones like the strophoid and the cissoid). Beyond conics the simplest such curve has equation

$$(x^2+y^2)L(x,y)^2-P(x,y)^2=0,$$

where $L(x,y)=0$ is the equation of a line and $P(x,y)=0$ is the equation of a conic.

Finally, given a conic C and a point m, let's see under what conditions the pedal of m with respect to C is a circle or a line. We saw at the end of section 3 that this pedal P is the inverse, under the inversion with pole m and power 1, of the polar transform C^0 of C with respect to the circle S of center m and radius 1. Thus the pedal P is a circle or a line if and only if C^0 is (section 1.6). This means that C^0 goes through the cyclic points. But the polars of cyclic points with respect to S are the isotropic lines at m, which are thus tangent to C. Hence:

Theorem 55. *The pedal P of a point m with respect to a proper conic C is a circle or a line if and only if m is a focus of C.* □

Saying that P is a line is the same as saying that $m \in C^0$, which means that C is tangent to the polar of m with respect to S, which is the line at infinity. Thus:

Corollary. *The pedal of m with respect to a conic C is a line if and only if C is a parabola and m its focus.* □

In general, if we make m the origin and write the equation of C^0 in the form $F_2(x,y) + F_1(x,y) + k = 0$, with F_i homogeneous of degree i, the equation of the pedal P is obtained by replacing x by $x/(x^2+y^2)$ and y by $y/(x^2+y^2)$ (formula (25) in section 1.6). Thus we obtain

$$F_2(x,y) + (x^2+y^2)F_1(x,y) + k(x^2+y^2)^2 = 0. \tag{72}$$

If $k \neq 0$ and m is not a focus of C, this is the equation of a quartic having the cyclic points as double points (a *bicircular quartic*). If $k = 0$, which means that C^0 goes through m and C is a parabola, we obtain a *circular cubic* having m as a double point. This cubic is cuspidal (section 2.6) if and only if F_2 is a square, which means that C^0 is a parabola, that is, $m \in C$; otherwise it is nodal, and is a strophoid if and only if C^0 is an equilateral hyperbola (the tangents to C drawn from m are orthogonal).

Since C is unicursal, so is its pedal.

Appendix: (2, 2)-Correspondences

A *correspondence* between two sets D and D' is a subset B of the product $D \times D'$. To a point $a \in D$ correspond all the points in $\mathrm{proj}_{D'}((a \times D') \cap B) \subset D'$, that is, all points $a' \in D'$ such that $(a, a') \in B$; and conversely.

Here D and D' will be algebraic varieties contained in the projective spaces $\mathbf{P}$ and $\mathbf{P}'$, and their product is a subset of $\mathbf{P} \times \mathbf{P}'$, which in turn is embedded as a Segre variety (section 3.3) in a projective space $\mathbf{P}''$. Even if D and D' are simply projective lines, we'll be doing algebraic geometry in dimension three or more, and for that we will need some more sophisticated results than we have used so far. These few results, stated below, will not be demonstrated.

Background results

(a) The subsets of a projective space $\mathbf{P}_n$ defined by systems of homogeneous polynomial equations $F_j(x) = 0$ are called *algebraic*. Finite unions and arbitrary intersections of algebraic sets are algebraic. The set of algebraic subsets of $\mathbf{P}_n$, ordered by inclusion, has the property that every non-empty subset has a minimal element; in other words, every decreasing sequence of algebraic sets is stationary. We call *irreducible* a non-empty algebraic set A that is not the union of two algebraic sets A' and A'' distinct from A; the existence of a minimal element implies that every algebraic set B is a finite union of irreducible subsets. An *(algebraic) variety* is an irreducible algebraic set. A decomposition $B = V_1 \cap \cdots \cap V_q$ of B as the union of varieties V_i none of which is contained in another is unique, and the V_i are called the *(irreducible) components* of B.

(b) By various algebraic and analytic methods, one can assign to every algebraic variety V an integer $\dim V$, called its *dimension*. This assignent generalizes the notion of dimension for projective linear spaces (section 1.1). Varieties of dimension zero are points, if we assume, as we do from now on, that the field of scalars is algebraically closed.

(c) An algebraic subset of $S \subset \mathbf{P}_n$ all of whose components are $(n-1)$-dimensional is called a *hypersurface*, or a *surface* if $n = 3$. Hypersurfaces are exactly the algebraic subsets that can be defined by a single homogeneous polynomial equation $F(x_0, \ldots, x_n) = 0$ (cf. section 1.3). If we assume that F has no repeated factors (and so is determined by S up to a constant multiplicative factor, cf. section 1.8), S is a variety if and only if F is irreducible; the degree of F is called the *degree* of the hypersurface S, and denoted by $d^0(S)$. This degree is the number of intersections of S with an arbitrary line not contained in S, counted with their multiplicities.

(d) An algebraic set all of whose components have dimension 1 is called a *curve*. The intersection $S \cap S'$ of two surfaces in $\mathbf{P}_3$ without common components is a curve. One can assign a multiplicity $m(i)$, or, more explicity, $m(C_i; S \cdot S')$ to each component C_i $(i = 1, \ldots, q)$ of this intersection; the formal sum $m(1)C_1 + \cdots + m(q)C_q$, a *1-cycle*, is called the *intersection cycle* of S and S', and is denoted by $S \cdot S'$. The condition $m(i) = 1$ is equivalent to the existence of a point of C_i where the tangent planes to S and S' are distinct.

Conversely, every irreducible curve C is a component of the intersection of two surfaces. If it is actually equals the intersection of some two surfaces S and S', that is, if it coincides with the cycle $S \cdot S'$, where $m(C; S \cdot S') = 1$, we say the curve is a *complete intersection*; not every curve is a complete intersection.

(e) Given a curve C and a surface S in $\mathbf{P}_3$ that does not contain any component of C, the intersection $S \cap C$ is made up of a finite number of points P_j, for $j = 1, \ldots, r$. Each point can be assigned a multiplicity $m(j)$ or $m(P_j; C \cdot S)$. The number of such points, counted with multiplicities, is a multiple $nd^0(S)$ of the degree of S; the number n, which only depends on C, is called the *degree* of C and denoted by $d^0(C)$. The curve C is a line if and only if $d^0(C) = 1$; a curve C intersects a plane in $d^0(C)$ points. The sum of degrees of components of the intersection $S \cdot S'$ of two surfaces, counted with multiplicities, is $d^0(S)d^0(S')$.

By linearity, this generalizes to a divisor S and a one-cycle C.

The formal sum $m(1)P_1 + \cdots + m(r)P_r$, a *0-cycle*, is denoted by $S \cdot C$ and called an *intersection cycle*.

(f) We shall admit that all (reasonable) ways of computing intersection multiplicities (for example, as the multiplicities of roots of polynomial equations) yield the same result.

Given three surfaces (or divisors) S, S' and S'', the associativity formula $(S \cdot S') \cdot S'' = S \cdot (S' \cdot S'')$ holds, as long as both sides are defined.

(g) Let P be a point in the intersection C of two surfaces S and S'. Then P is *simple* on C if and only if it is simple on S and S' and the planes tangent to S and S' at P are distinct. Thus the intersection of two surfaces tangent at a point P (for example, a surface and its tangent plane) has P as a multiple point.

For a surface and its tangent plane, this follows from a very simple calculation.

Since the intersection of a quadric and a plane is a conic, the intersection of a quadric and a plane tangent to it is a pair of lines.

Correspondences between two projective lines

Let D and D' be projective lines. We will limit ourselves to correspondences that are curves (or 1-cycles) in $D \times D'$. By taking projective coordinate ratios u and v on D and D', such a correspondence C is described by a polynomial relation

(a) $$F(u, v) = 0.$$

If p and q are the degrees of F with respect to u and v, respectively, we say that C is a *(p, q)-correspondence*. Notice that (a) does not define a curve in the affine plane K^2, nor even in $\mathbf{P}_2(K)$, because u and v can be infinity.

Recall that $\mathbf{P}_1 \times \mathbf{P}_1$ and $\mathbf{P}_2$ are not isomorphic. On $\mathbf{R}$, one is a torus, the other a non-orientable surface (section 2.8). Over a finite field, these two varieties have different cardinalities, namely $(q+1)^2$ and q^2+q+1.

To be rigorous, one should really take homogeneous coordinates (u', u'') on D (so $u = u'/u''$) and (v', v'') on D' (so $v = v'/v''$). With this notation $u''^p v''^q F(u'/u'', v'/v'')$ is a polynomial $F_h(u', u''; v', v'')$, homogeneous of degree p in (u', u'') and degree q in (v', v''). But since we are already accustomed with working with ∞ (considering infinite roots of equations, for instance) we need not resort to this heavier notation.

The points of D' in correspondence with a point $u_0 \in D$ are simply the roots (finite or not, distinct or not) of the equation $F(u_0, v) = 0$. If C is a (p, q)-correspondence, there are q such roots. Similarly, there are p points of D in correspondence with a point $v_0 \in D'$.

A correspondence C is said to be *decomposable* if it is not an irreducible curve. It is *degenerate* if one or more of the horizontal or vertical lines, $D \times v_0$ or $u_0 \times D'$, appear in the decomposition. A $(p, 0)$-correspondence is made up of p vertical lines; if a point $u_0 \in D$ is not on the projection of these lines, it has no corresponding point in D'.

For a $(1,1)$-correspondence, formula (a) can be written $auv+bu+cv+d = 0$, which is a projective transformation, unless $ad - bc = 0$. If we represent

$D \times D'$ by the Segre quadric S, with equation $xt - yz = 0$, where $x = u'v'$, $y = u'v''$, $z = u''v'$ and $t = u''v''$ (section 3.3), the homogeneous equation of the correspondence,

$$au'v' + bu'v'' + cu''v' + du''v'' = 0,$$

reduces to $ax+by+cz+dt = 0$, so the correspondence C is the intersection of S with a plane, hence a conic. (1, 1)-correspondences such that $ad - bc = 0$ are the intersections of S with its tangent planes; they decompose into the two generators at the contact point, namely, a vertical line and a horizontal one.

A (2, 1)-correspondence can be written $(au^2+bu+c)v+(a'u^2+b'u+c') = 0$. If it is irreducible, it is a rational map of degree two from D onto D'. Its homogeneous equation is

$$(au'^2 + bu'u'' + cu''^2)v + (a'u'^2 + b'u'u'' + c'u''^2)v'' = 0.$$

The left-hand side is not a function of the products $x = u'v'$, etc., only, so C is not the intersection of S with another surface T; this could be expected because, if it were, the elements of each of the two families of S would intersect T in the same number $d^0(T)$ of points. More generally, this reasoning, together with a calculation similar to the one made for projective transformations, show that:

Theorem A. *A correspondence is a(complete) intersection of the Segre quadric S with another surface T if and only if it is (n, n), where $n = d^0(T)$.* □

An irreducible (2,1)-correspondence C intersects a plane tangent to S at three points (two on a generator, one on another). Thus such a correspondence has degree three. It is not a plane curve, otherwise it would be the intersection of S with a plane, hence a conic; we say it is a *twisted cubic*. The equation $P(u)v + Q(u) = 0$, with P and Q quadratic polynomials, and the parametric representation $x = uv$, $y = u$, $z = v$, $t = 1$ of S show that C is a unicursal cubic with homogeneous parametric representation $\big(-uQ(u), uP(u), -Q(u), P(u)\big)$; this shows again that the degree of the curve is three. Its representation is proper.

A twisted cubic cannot be the complete intersection of two surfaces T and T', for the relation $d^0(T)d^0(T') = 3$ would force either T or T' to be a plane.

The union of C with a horizontal generator G, for instance, $v = 0$, is a complete intersection of S with a surface T (to see this, multiply the homogeneous equation of C by v'). This union is a degenerate (2, 2)-correspondence.

(2, 2)-correspondences and biquadratics; degenerate cases

Let B be a (2, 2)-correspondence between two projective lines D and D'. Let

(b) $\quad au^2v^2 + uv(bu + b'v) + cuv + c'u^2 + c''v^2 + du + d'v + e = 0$

be its equation, or, in homogeneous form,

$$au'^2v'^2 + u'v'(bu'v'' + b'v'u'') + cu'v'u''v'' + c'u'^2v''^2 + c''v'^2u''^2 + du'u''v''^2 + d'v'v''u''^2 + eu''^2v''^2 = 0.$$

This is the intersection of the Segre quadric S with the quadric T with equation

(c) $$ax^2 + x(by + b'z) + cxt + c'y^2 + c''z^2 + dyt + d'zt + et^2 = 0.$$

The equation of T is only determined modulo the equation $xt - yz$ of S: one could have written cyz, or $c_1yz + (c - c_1)xt$, instead of cxt. Thus the quadric T is the most general possible. Whence:

Theorem B. *Every (2, 2)-correspondence is the intersection of the Segre quadric S with an arbitrary quadric $T \neq S$. Every quadric T' in the pencil determined by S and T gives rise to the same correspondence as T.* □

The curve $T \cdot S$ is called a *biquadratic* of S. It intersects an arbitrary plane P in four points (distinct or not), namely the points common to the conics $P \cdot S$ and $P \cdot T$. Thus a biquadratic has degree four. Its projections on planes of $\mathbf{P}_3$ are, in general, quartics.

The following notations will be useful in describing and studying the degenerate cases of biquadratics:

- B is an irreducible biquadratic;
- C and C' are irreducible twisted cubics, coming from (1, 2)- and (2, 1)-correspondences, respectively;
- H and H' are irreducible conics, coming from projective transformations;
- G_i and G'_i are vertical (u = constant) and horizontal (v = constant) generators, respectively.

The degenerate cases of a (2, 2)-correspondence E are the following:

(a) $E = H + H'$. This means that the pencil of quadrics determined by S and T contains the sum $P + P'$ of two planes; their common line $P \cdot P'$ intersects S, and also T, at two points p, p' where S and T are tangent (since their tangent planes at p and p' contain the tangents to H and H'). Conversely, if S and T are tangent at two points p and p', their intersections with any plane P going through p and p' are bitangent at these points, and they can have no other common point unless they coincide; since $E = S \cdot T$ contains common points other than p and p', the intersections of S and T with at least one plane P coincide, and S and T share a conic H. The rest of E is a conic H'.

(b) $E = 2H$. This means that the pencil contains a double plane, the plane of H. Thus S and T are tangent at all points of H. Conversely, if S and T are tangent along a common conic, we're in this case.

(c) $E = C + G$. This means that S and T only have one common line G, which intersects the cubic C in two points (distinct or not), since we're dealing with a $(1,2)$-correspondence; thus, in general, S and T are tangent at two points of a same generator G of S. Conversely, G is then bitangent to T and hence contained in T.

(c') $E = C' + G'$. This case is symmetric with the previous one.

(d) $E = H + G + G'$. This means that the pencil determined by S and T contains the sum of a plane tangent to S (at the intersection point of G and G') with another plane.

(e) $E = G_1 + G_2 + G_1' + G_2'$. This means that the pencil determined by S and T contains the sum of two planes tangent to S, and this in two different ways. The reduced form of its general equation is $xt - qyz = 0$, for $q \in \hat{K}$.

Symmetric and symmetrizable correspondences

From now on we assume that K has characteristic $\neq 2$ (recall that we already have assumed that K is algebraically closed).

Let C be a correspondence between a set D and itself. One says that C is *symmetric* if the relations $(p,q) \in C$ and $(q,p) \in C$ are equivalent. If D is a projective line and C is an algebraic correspondence with equation $F(u,v) = 0$, this means that $F(u,v) = kF(v,u)$ (cf. section 1.8). We deduce that $k^2 = 1$, that is, $k = \pm 1$. Here we will restrict the definition of a *symmetric correspondence* to one such that $F(u,v) = F(v,u)$, that is, when the polynomial $F(u,v)$ is symmetric in u and v (and, consequently, is a polynomial in $u+v$ and uv).

Thus the identity $u - v = 0$ is not a symmetric correspondence, rather a skew-symmetric one ($k = -1$). For a skew-symmetric polynomial $F(u,v) = -F(v,u)$, we have $F(u,u) = 0$, so $F(u,v)$ is a multiple of $u - v$. A skew-symmetric $(2,2)$-correspondence is thus the sum of the identity with an involution on D.

The degrees of a symmetric (or skew-symmetric) correspondence with respect to u and v are the same; thus such a correspondence is (n,n), and is a complete intersection with the Segre quadric S (theorem A).

A symmetric $(1,1)$-correspondence has an equation of the form $auv + b(u+v) + c = 0$; if it is not degenerate, it is an involution. The general equation of a symmetric $(2,2)$-correspondence is

$$\text{(d)} \qquad au^2v^2 + buv(u+v) + cuv + c'(u^2+v^2) + d(u+v) + e = 0.$$

This symmetry property is intrinsic: it is preserved if we apply the same homographic transformation to u and v.

The notion of a symmetric correspondence does not make sense on the product $D \times D'$ of two different projective lines. However, the following result is interesting and useful:

Theorem C. *Let B be a $(2,2)$-correspondence between two projective lines D and D'. Except for the obvious cases when B is degenerate, there exists*

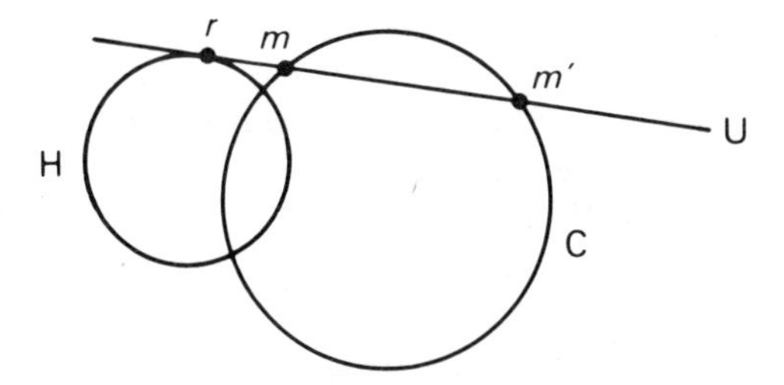

at least one projective transformation $h : D \to D'$ such that the correspondence B' on $D \times D$ defined by "$(p, q) \in B'$ if and only if $(p, h(q)) \in B$" is symmetric.

In other words, one can symmetrize the equation $F(u, v) = 0$ by applying a projective transformation to one of the projective coordinate ratios.

Proof. Let B be the intersection of the Segre quadric S with another quadric T'. In the linear pencil of quadrics determined by S and T', there is at least one quadric T whose equation is a degenerate quadratic form, that is, a cone: indeed, if we write the general equation of the pencil as $Q+rQ' = 0$, where Q and Q' are quadratic forms and $r \in \hat{K}$, the degeneracy of $Q + rQ'$ corresponds to the vanishing of an determinant of order four, which gives a degree-four equation in r. In general there are four cones (distinct or not) in the pencil.

Let s be the vertex of such a cone T. We will apply a *projection with center* s. The contact points with S of the lines going through s and tangent to S describe a conic H, the intersection of S with the polar plane P of s (theorem 49 in section 4.1); thus these tangents form a cone of degree two.

H is what is called the *apparent contour* of the projection of S seen from s.

A generator G of S has as its projection on P the tangent to H at the point $G \cdot P = G \cdot H$; thus two generators G, G' belonging to different families have the same projection U on P if and only if their intersection is on H.

Now let C be the intersection of the plane P with the cone T. The intersections of the generator G (resp. G') with the biquadratic B, when projected to P, give the intersections m, m' of C with the tangent U to H at $G \cdot P$ (resp. $G' \cdot P$). The corresponding points of $G \cdot B$ (resp. $G' \cdot B$) are $G \cdot D_{ms}$ and $G \cdot D_{m's}$ (resp. $G' \cdot D_{ms}$ and $G' \cdot D_{m's}$).

Let h be the projective transformation from D onto D' whose graph is H. For $p \in D$ and $q \in D'$, the relation $(p, h(q)) \in B$ says that the vertical generator G_p and the horizontal generator $G'_{h(q)}$ intersect on B. By the definition of h, $G'_{h(q)}$ is the horizontal generator that intersects G_q on H; similarly, $G'_{h(p)}$ is the vertical generator that intersects G_p on H. Since the points m and m' are the same for G_q and $G'_{h(q)}$, and also for G_p and $G'_{h(p)}$, the relations $(p, h(q)) \in B$ and $(q, h(p)) \in B$ are equivalent. Whence the symmetry of B'.

This reasoning assumes that at least one of the cones T of the pencil is non-degenerate. We now look at the case when something funny is going on: the vertex of the cone is on S, or the cone splits into two planes. Assume first that B is irreducible, which in particular excludes the possibility of T splitting into two planes. If the vertex s of T is on S, all the other quadrics of the pencil have the same tangent plane at s, and s is a double point of B. The polar P of s is the tangent plane, H splits into two generators of S and does not furnish a true projective transformation. Placing this double point at $u = 0$, $v = 0$, we get $d = d' = e = 0$ in equation (b), which then becomes

$$\text{(e)} \qquad au^2v^2 + bu^2v + b'uv^2 + cuv + c'u^2 + c''v^2 = 0.$$

In order to preserve the condition $v = 0$, we try a projective transformation replacing v by $v/(pv + q)$, with $q \neq 0$. Then (e) becomes

$$au^2v^2 + bu^2v(pv + q) + b'uv^2 + cuv(pv + q) + c'u^2(pv + q)^2 + c''v^2 = 0.$$

The symmetry conditions are $bq + 2c'pq = b' + cp$ and $c'q^2 = c''$. Since $c' \neq 0$ and $c'' \neq 0$ (otherwise u or v is a factor in (e) and B is degenerate), the equation $c'q^2 = c''$ gives two non-zero values for q, one the negative of the other. The other equation, which can be written $(2c'q - c)p = b' - bq$, gives at least one value for p, for the two possible values for q don't both make $2c'q - c$ vanish, since $c' \neq 0$.

The relation $2c'q = c$ implies $4c'^2q^2 = c^2$, hence $4c'c'' = c^2$, which means that the double point of B is a cusp.

We now look at the case when B is degenerate. If B is the sum of two projective transformations k, k' from D onto D', we can symmetrize both simultaneously: write $k^{-1}k'$ as the product jj' of two involutions of D (theorem 31 in section 2.4), and set $h = k'j' = kj$. This projective transformation responds to the question. Things are even simpler when $B = 2H$. When $B = H + G + G'$, we take $u = 0$ for G, $v = 0$ for G'; the projective transformation can be written $auv + bu + cv + d = 0$. Applying a homothety to v makes this symmetric if b and c are both zero or both non-zero. If $c = 0$ and $b \neq 0$, say, we apply the projective transformation $v \to v/(v+q)$ instead, with $q \neq 0$; this gives $auv + bu(v+q) + d(v+q) = 0$, which is symmetric if we take $q = d/b$, which is non-zero because otherwise $d = 0$ and the projective transformation is degenerate. Finally, if $B = G_1 + G_2 + G_1' + G_2'$, the G_i and G_i' being simultaneously different or the same, we can just apply a projective transformation to u so as to give G_1 and G_1' the same value of u, and similarly for G_2 and G_2'.

In the other cases, $B = C + G$, $B = C' + G'$, $B = 2G + G_1' + G_2'$ and $B = G_1 + G_2 + 2G'$, the correspondence is obviously not symmetric. □

Critical points

Let B be a (2,2)-correspondence on $D \times D'$. A point $u \in D$ is said to be *critical* if its two corresponding points in D', namely those of $(u \times D') \cdot B$, are identical. This can be expressed by writing that (*b*), as an equation in v, that is,

$$\text{(b}'\text{)} \qquad (au^2 + b'u + c'')v^2 + (bu^2 + cu + d')v + c'u^2 + du + e = 0,$$

has a (finite or infinite) double root, which means that

$$\text{(f)} \qquad (bu^2 + cu + d')^2 - 4(au^2 + b'u + c'')(c'u^2 + du + e) = 0.$$

This is a degree-four equation in u. Thus there are four critical points on D (counting multiplicities); their formal sum, denoted by $\mathbf{d}$, is called the *critical divisor* of B on D. We say that the critical divisor is undefined if (f) is satisfied for any value of u.

Similar definitions apply to D', whose critical divisor is denoted by $\mathbf{d}'$.

This notion corresponds to that of the discriminant in number theory. Since the critical divisor is formed by the projections of the points of B where the tangent is vertical (or of multiple points of B), we can say that it is a kind of apparent contour.

The critical divisors of a symmetric correspondence on $D \times D$ are, of course, equal. Thus theorem C implies that:

Theorem D. *Except for obviously non-symmetrizable degenerate correspondences, there exists a projective transformation $h : D \to D'$ taking* $\mathbf{d}$ *to* $\mathbf{d}'$. □

If the four roots of (f) are distinct, their cross-ratio is equal to the cross-ratio of the corresponding equation in v (if we consider them in the right order).

The *type* of a divisor of degree 4 is the sequence of coefficients of its points: $(1,1,1,1)$, $(1,1,2)$, $(1,3)$, $(2,2)$ and (4). Except for the obvious degenerate exceptions, the critical divisors on D and D' of a correspondence have the same type, by theorem D. Here are some useful conclusions in the study of types:

(1) If B is degenerate and contains the vertical $u \times D'$, u appears in $\mathbf{d}$ with coefficient two or more (generally two). We see this by taking $u = 0$; then u is a factor in (b′), so $c'' = d' = e = 0$ and 0 is a multiple root of (f).

(2) If B has a double point (u, v), then u appears in $\mathbf{d}$ with coefficient two if B has distinct tangents there, and with coefficient ≥ 3 if the tangents coincide. To see this, take $u = v = 0$; then $d = d' = e = 0$ and equation (f) becomes $(bu+c)^2u^2 - 4c'(au^2+b'u+c'')u^2 = 0$. Its root 0 is double if $c^2 - 4c'c'' \neq 0$, triple at least if $c^2 - 4c'c'' = 0$, quadruple if we also have $bc - 2b'c' = 0$. The correspondence B can be irreducible in the

case of a triple root. If, on the other hand, $c^2 - 4c'c'' = 0$ and $bc = 2b'c'$, we have two possibilities: either $c' = c'' = 0$ and B is non-degenerate, or (taking $c' = 1$) $c'' = c^2/4$ and $b' = \frac{1}{2}bc$, in which case (b′) can be written

$$au^2v^2 + buv(u + \tfrac{1}{2}cv) + (u + \tfrac{1}{2}cv)^2 = 0;$$

then B splits into two projective transformations, whose graphs are tangent at $(0, 0)$.

(3) If B is of the form $C + G$, we can assume that $(0, 0)$ is on $C \cdot G$, so that $c'' = d' = e = d = 0$ and that (f) is of the form

$$(bu^2 + cu)^2 - 4(au^2 + b'u)c'u^2 = 0.$$

The root $u = 0$ is triple if and only if $c = 0$, which means that C, whose equation is $auv^2 + v(bu + b'v) + cv + c'u = 0$, is tangent to G ($u = 0$) instead of intersecting it at two points; but this root cannot be quadruple, otherwise $b'c' = 0$ and the equation of C would have u or v as a factor. On the other hand, the critical divisor of $B = C + G$ on D', given by the discriminant $(b'v^2 + cv)^2 = 0$ of the equation (in u) $(av^2 + bv + c')u^2 + (b'v^2 + cv_u = 0)$, is of type $(2, 2)$ if $c \neq 0$ or (4) if $c = 0$.

(4) Conversely, if u appears with coefficient ≥ 2 in $\mathbf{d}$, we can assume that the critical point is $u = 0$ and the corresponding value of v is $v = 0$; then $d' = e = 0$ and (f) is of the form

$$(bu + c)^2u^2 - 4(au^2 + b'u + c'')(c'u + d)u = 0.$$

This equation has 0 as a multiple root if and only if $c''d = 0$. If $c'' = 0$, u factors out in (b′) and B contains $0 \times D'$; if $d = 0$, $(0, 0)$ is a multiple point of B.

These observations allow us to explicitly write the type of the critical divisor(s) as a function of the geometry of the given (2,2)-correspondence:

B without multiple point	$(1, 1, 1, 1)$
B with node	$(2, 1, 1)$
B with cusp	$(3, 1)$
$C + G$, with C not tangent to G	$(2, 1, 1)$ on D, $(2, 2)$ on D'
$C + G$, with C tangent to G	$(3, 1)$ on D, (4) on D'
$C' + G'$, with C' not tangent to G'	$(2, 2)$ on D, $(2, 1, 1)$ on D'
$C' + G'$, with C' tangent to G'	(4) on D, $(3, 1)$ on D'
$H + H'$ with two common points	$(2, 2)$
$H + H'$ with tangency	(4)
$2H$	undefined
$H + G + G'$ with $G \cdot G' \notin H$	$(2, 2)$
$H + G + G'$ with $G \cdot G' \in H$	(4)
$G_1 + G_2 + G'_1 + G'_2$ (distinct)	$(2, 2)$
$2G + 2G'$	undefined
$G_1 + G_2 + 2G'$	undefined on D, (4) on D'
$2G + G'_1 + G'_2$	(4) on D, undefined on D'

Examination of this table shows that:

Theorem E. *The geometry of a non-degenerate* $(2,2)$*-correspondence is uniquely determined by the type of its critical divisor. If we allow also symmetrizable degenerate correspondences, only types* $(2,2)$ *and* (4) *are ambiguous.* □

Geometric interpretation of $(2,2)$*-correspondences*

Let D be a proper conic, with its canonical projective line structure (section 2.5). One example of a $(2,2)$-correspondence on $D \times D$ is the following: take an envelope of second class E and say that two points $m, m' \in D$ are in correspondence if and only if $D_{mm'} \in E$. Indeed, every point $m \in D$ lies in two lines of the family E, and the two points where these lines intersect D again are the points that correspond to m.

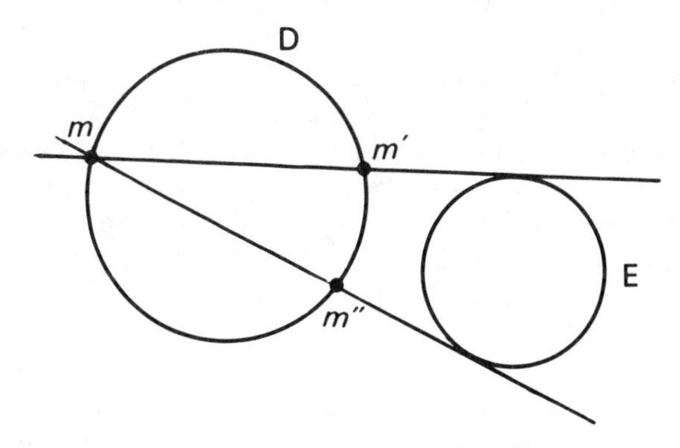

Conversely, we have:

Theorem F. *Every symmetric* $(2,2)$*-correspondence on a proper conic* D *is defined by an envelope of second class* E*, and can be written* $D_{mm'} \in E$.

Proof. In an appropriate projective frame, we can take the parametric representation of D to be $x = t$, $y = t^2$, $z = 1$. The line $D_{mm'}$ joining the points m, m' with parameters t and t' of D has equation $-(t+t')x + y + tt'z = 0$. Let $F(u,v,w)$ be the tangential equation of E, having the explicit form

$$Lu^2 + L'v^2 + L''w^2 + Mvw + M'wu + M''uv = 0.$$

The relation $D_{mm'} \in E$ becomes $F\bigl(-(t+t'), 1, tt'\bigr) = 0$, that is,

$$L(t+t')^2 + L' + L''t^2t'^2 + Mtt' - M'tt'(t+t') - M''(t+t') = 0.$$

This can be identified with the general symmetric $(2,2)$-correspondence

$$at^2t'^2 + btt'(t+t') + ctt' + c'(t^2+t'^2) + d(t+t') + e = 0$$

by taking $L = c'$, $L' = e$, $L'' = a$, $M = c - 2c'$, $M' = -b$, $M'' = -d$. □

When the envelope E splits into two points p and p', the correspondence B is the sum $H + H'$ of the Frégier involutions (theorem 35 in section 2.5) defined by p and p', if $p, p' \notin D$. If $p' \in D$ but $p \notin D$, B is of the form $H + G + G'$. If $p, p' \in D$, B splits into four generators.

When E is a double point $2p$ we have $B = 2H$ if $p \notin D$ and $B = 2G + 2G'$ if $p \in D$.

From now on we asssume that the envelope E is non-degenerate. Then E is formed by the tangents to a proper conic, also denoted by E.

We could instead take the dual point of view, where two points m, m' of a conic D are in correspondence if and only if the tangents to D at m and m' intersect on another conic E'.

We now introduce two important notions.

A point $m \in D$ is *fixed* under a correspondence B if it is in correspondence with itself, that is, if $(m, m) \in B$. Fixed point can be found in several ways:

- By making $t = t'$ in the equation of B. This gives a degree-four equation and four fixed points, or rather a degree-four divisor $\mathbf{f}$.
- By intersecting B with the diagonal I. Since I is a plane section of the *Segre quadric* and B is a curve of degree four, $B \cdot I$ is a degree-four divisor. We have admitted that its projection on D is $\mathbf{f}$.
- By looking for points $m \in D$ such that the tangent to D at m is also tangent to E. In other words, fixed points correspond to *common tangents* to D and E, and we know that there are four such tangents. The points corresponding to m are m itself and the other point m' where the other tangent to E passing through m intersects D again; we have $m' = m$ if and only if $m \in E$, that is, if and only if D and E are tangent at m, and then m counts as two fixed points at least.

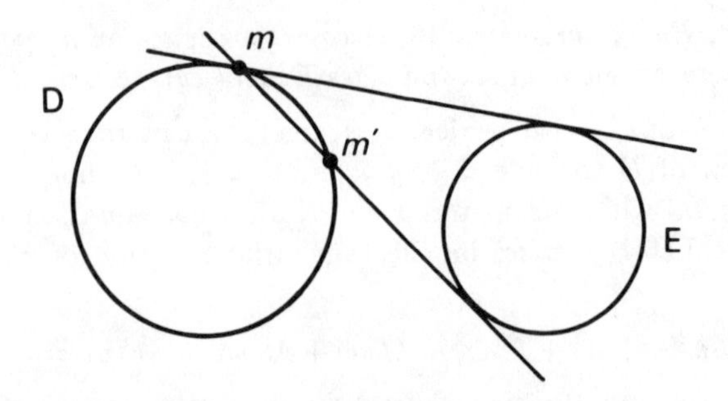

A point $a \in D$ is a *critical point* if and only if the two tangents to E passing through a coincide, that is, if and only if a is on E. This suggests the following result:

Theorem G. *With the notations above, the critical divisor* $\mathbf{d}$ *on* D *is equal to the intersection cycle* $D \cdot E$.

Proof. This is obvious if D and E have four distinct common points, and "all hell would break loose" if in other cases the points of $D \cdot E$ occurred with different coefficients in $\mathbf{d}$. Even then, let's write out a proof. Let $F(u,v,w) = 0$ be the tangential equation of E. The line (u,v,w) goes through (x,y,z) if and only if $ux + vy + wz = 0$. Then $F\big(-(vy + wz), vx, wx\big)$ is a quadratic homogeneous polynomial in v,w, whose discriminant $F^0(x,y,z)$ is the left-hand side of the punctual equation of E (section 4.3). As in theorem F, we replace x,y,z by $(t,t^2,1)$. Then $F^0(t,t^2,1)$ is the discriminant of the homogeneous polynomial

$$\big(F\big(-(vt^2 + w), vt, wt\big),$$

and also the discriminant of the ordinary polynomial $F\big(-(t^2+W), t, tW\big)$, where $W = w/v$, and that of $F\big(-(t+T), 1, tT\big)$, where $W = tT$. But we saw in theorem F that $F\big(-(t+t'), 1, tt'\big) = 0$ is the equation of the correspondence B, so $F^0(t^2,t,1) = 0$ is indeed its critical divisor. □

We have just found a necessary condition for the containment of hell (but not a sufficient one!).

The table preceding theorem E then shows that:

Corollary.

- (1,1,1,1) *B is a biquadratic without multiple points if and only if D and E have four distinct common points;*
- (2,1,1) *B is a biquadratic with one node if and only if D and E are tangent at one point;*
- (3,1) *B is a biquadratic with a cusp if and only if D and E osculate;*
- (2,2) *B is the sum $H + H'$ of two projective transformations having two common points if and only if D and E are bitangent;*
- (4) *B is the sum $H + H'$ of two projective transformations having a common tangent if and only if D and E superosculate.* □

We will now make a study, due to Poncelet, of polygons inscribed in a conic D and circumscribed to another conic E. The idea is to use the composition of a $(2,2)$-correspondence on D with itself. This $(2,2)$-correspondence B on D is defined as the set of pairs (m,m') such that $D_{mm'}$ is tangent to E. The composition can be visualized in the following way: through a point m_0 of D we consider the tangents to E; they intersect D again at points m_1 and m_{-1}. Through m_1 we now draw the other tangent to E (distinct from $D_{m_0m_1}$), which intersects D again at m_2, and from m_{-1} we derive m_{-2} by the same procedure. We continue in this way for $n = 3,4,\ldots$, associating to m_0 points m_n, m_{-n} that are rational functions of m_1 and m_{-1}; but m_1 and m_{-1} themselves are quadratic in $K(m_0)$,

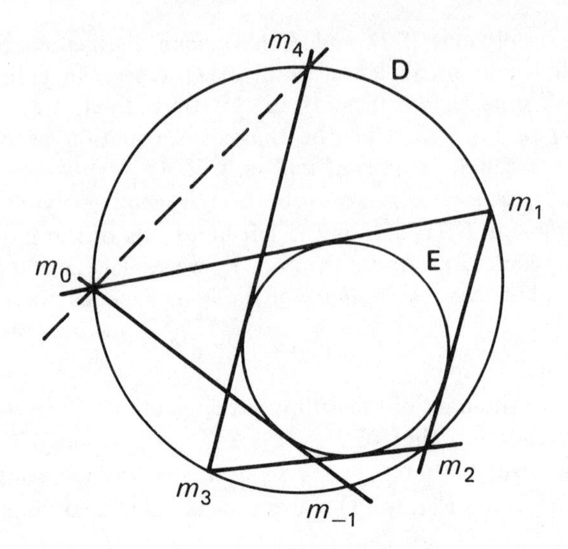

so the pairs (m_0, m_n) and (m_0, m_{-n}) describe an algebraic correspondence B_{n+1} on D.

Since the same procedure can take us backwards from m_n to m_0, the correspondence B_{n+1} is symmetric, which means that it is determined on D by the lines in an envelope E_{n+1} of second class. By theorem F, this envelope is formed, in general, by the tangents to a proper conic E_{n+1}. In other words, as m_0 describes D the lines $D_{m_0 m_n}$ and $D_{m_0 m_{-n}}$ form, in general, the envelope of a conic E_{n+1}.

Theorem H (Poncelet). *If the conic E_{n+1} is proper, it belongs to the punctual pencil determined by D and E(at least when they have four distinct common points).*

Proof. Indeed, if m_0 is a critical point of B, m_1 and m_{-1} coincide. By recurrence, so do m_n and m_{-n}, which implies that m_0 is a critical point of B_{n+1}. The result now follows from theorem G, at least when the four points in the intersection of D and E are distinct. It would be pretty weird if things didn't work out in other cases—but they do. Roughly speaking, the reason is that algebraic relations are preserved when we give particular values to independent coefficients—here those of the tangential equation of a conic E (compare the proof of theorem G). □

Theorem I (Poncelet). *With the notations above, assume there exists a point $a_0 \in D$ and an integer $n \geq 3$ such that the polygon $(a_0, a_1, \ldots, a_{n-1}, a_0)$ is circumscribed to E(by which we mean that $a_{n-1}a_0$ is tangent to E), and that $D_{a_j a_0}$ is not tangent to E for any $2 \leq j \leq n-1$. Then, for every $m_0 \in D$, the polygon $(m_0, m_1, \ldots, m_{n-1}, m_0)$ is circumscribed to E, and in particular, $m_{-1} = m_{n-1}$, $m_{-2} = m_{n-2}$, and so on.*

Consider the correspondence B_{n+1}. We are supposed to show that it equals the identity I, or rather $2I$. Both of the points in correspondence with a_0 are a_0 itself (go around the polygon both ways), so a_0 is a double fixed point of B_{n+1} (this means that (a_0, a_0) appears with coefficient at least two in the intersection cycle $I \cdot B_{n+1}$, if this cycle is defined).

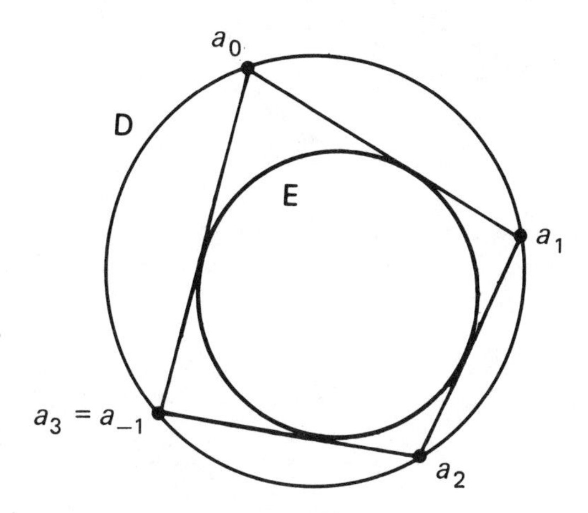

If we go around the polygon starting at a_j, for $j = 1, \ldots, n-1$, we see that all these points are double fixed points. By assumption, they are all distinct, so we have at least $2n \geq 6$ fixed points. Since B_{n+1} shares more than four points with the diagonal I, it must contain I; the rest of B_{n+1} is either a projective transformation or a sum of two generators, and it must contain the n points (a_j, a_j). If it is a projective transformation H we must have $H = I$, again by an excess of fixed points ($n > 2$); if it were a sum of two generators these n points would not be double fixed points of B_{n+1}, so this case is excluded. The only possibility, then, is $B_{n+1} = 2I$. □

There exists another proof based on critical points, but it assumes that $a_0 \notin D \cap E$. As we did above, we conclude that the a_j are critical points of B_{n+1}, and the critical divisor is then of degree $\geq 4 + 1$ (theorem G). Thus it is not defined, and, by the table preceding theorem E, B_{n+1} is of the form $2H$, where H is a projective transformation. But H has $n \geq 3$ fixed points $a_0, \ldots, a_{n-1}$, so it must be the identity.

We now spell out some particular cases of theorem I:

(a) Start from a point $m_0 \in D \cap E$. If D and E are tangent there, we stay at m_0 forever; otherwise m_1 and m_{-1} coincide with the point where the tangent to E at m_0 intersects D again. By induction, $m_j = m_{-j}$ for all j; the polygon closes up only if $D_{m_{n-1}m_0}$ is tangent to E, which means $m_{n-1} = m_1$, $m_{n-2} = m_2$ and so on.

If $n = 2q$ is even, the center region of the sequence $(m_0, \ldots, m_{2q-1})$ is $(\ldots, m_{q-1}, m_q, m_{q-1}, \ldots)$, and the polygon backtracks after bounc-

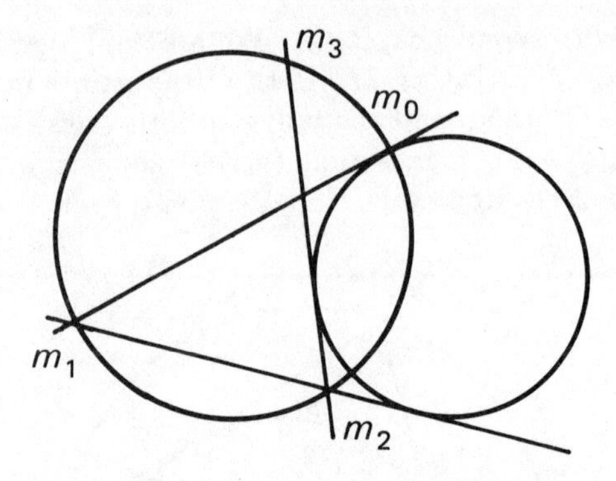

ing at m_q: the two tangents to E passing through m_q coincide with $D_{m_q m_{q-1}}$, and m_q is a point in the intersection of D and E (different from m_0, by assumption).

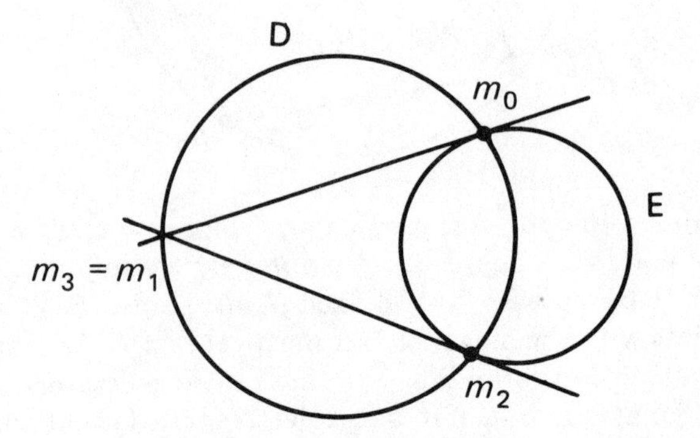

If $n = 2q + 1$ is odd, the center region of the sequence is $(\ldots, m_{q-1}, m_q, m_q, m_{q-1}, \ldots)$, so the tangent to D at m_q is also tangent to E. Here m_q is the contact point of D with one of the common tangent to D and E.

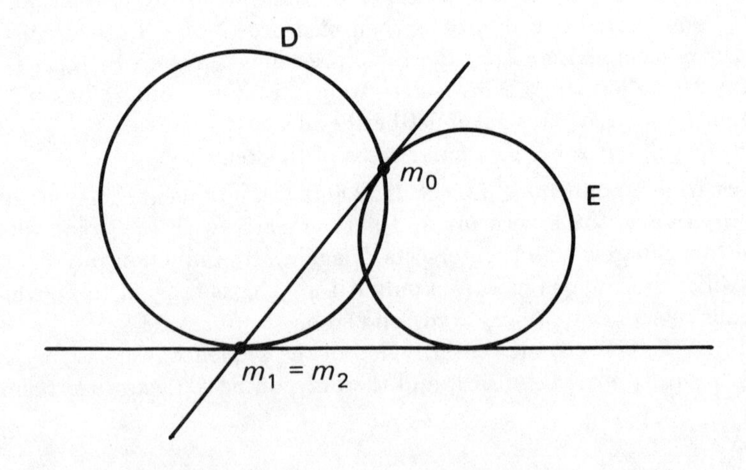

The two figures above illustrate the cases $n = 4$ and $n = 3$.

(b) If we start from a point m_0 lying on a tangent common to D and E, and if $m_1 \neq m_0$, which means that $m_0 \notin D \cap E$, we have $m_{n-1} = m_0$, $m_{n-2} = m_1$, and so on. As above, if $n = 2q + 1$ is odd, the polygon backtracks after bouncing at m_q, which is in the intersection $D \cap E$. If $n = 2q$ is even, the polygon backtracks after bouncing at m_q, which is a contact point of another common tangent.

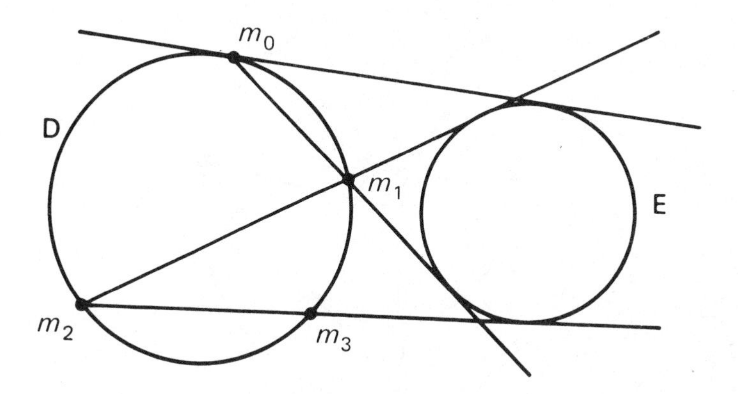

Notice that the sequence denoted by $(m_{-2}, m_{-1}, m_0, m_1, m_2)$ in the construction preceding theorem H has the following form (up to reflection):

- $(m_2, m_1, m_0, m_1, m_2)$ if $m_0 \in D \cap E$;
- $(m_1, m_0, m_0, m_1, m_2)$ if m_0 is the contact point of a common tangent.

(c) The conics D and E are bitangent or superosculating. We have seen (corollary to theorem G) that B is then of the form $H + H'$, where H and H' are projective transformations whose graphs have two common points or are tangent. Since $H + H'$ is symmetric, the two projective transformations are inverse to one another; we denote them by h and h^{-1}. They have the same fixed points, distinct or not, depending on whether D and E are bitangent or superosculating. Poncelet's theorem (theorem I) boils down to saying that $h^n = 1$; this is always true, for some n, if D and the B have their coefficients in a finite field.

When D and E are bitangent and we give the common points the values 0 and ∞, we can write $h(t) = ct$ (section 2.4), and Poncelet's theorem says that c is an n-th root of unity. When D and E are superosculating and we give the value ∞ to their common point, we get $h(t) = t + d$; then h never has finite order in characteristic zero, and always has order p in characteristic $p \neq 0$.

When D and E are bitangent in $\mathbf{P}_2(\mathbf{C})$, a projective transformation sends their common points to the cyclic points, so D and E become concentric circles (section 1.6). The projective transformations h and h^{-1} are rotations, with angles V and $-V$, around the common center (theorem 32 in section 2.4). If we denote by r and r' the radii of D and E, we have $\cos(\frac{1}{2}V) = r'/r$. The conclusion of Poncelet's theorem is true if and only if h has finite order n, that is, if and only if e^{2iV} is a primitive n-th root of unity; this gives an arithmetic relation between the radii r and r', namely, $r = 2r'$ for $n = 3$, $r^2 = 2r'^2$ for $n = 4$, $r'^2 - 2rr' - r^2 = 0$ for $n = 5$, $4r'^2 - 3r^2 = 0$ for $n = 6$, and so on.

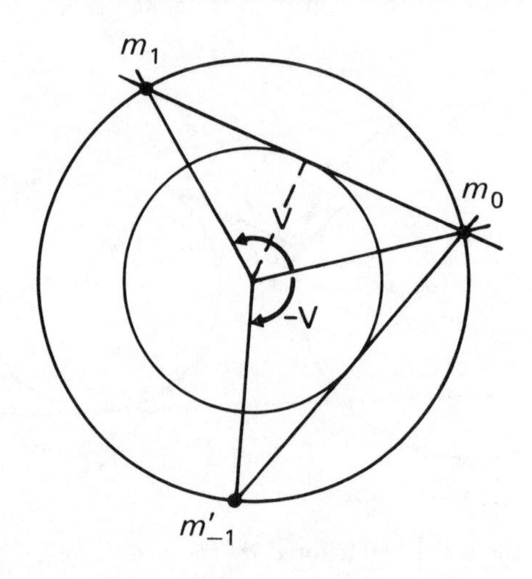

Curves on a quadric

We now consider a quadric S without double points, that is, one given by a quadratic form of maximal rank 4. Since the field is algebraically closed, the equation of such a quadric is just $xy - zt = 0$, and S can be seen as the product of two projective lines, whose projective coordinate ratios we denote by u and v.

A complete intersection of S with a surface T with homogeneous equation $P(x, y, z, t) = 0$ is an (n, n)-correspondence, for $n = d^0(T)$, whose equation is $F(u, v) = P(uv, u, v, 1) = 0$, where u and v range over $\hat{K}$ (theorem A). More generally, every curve (or 1-cycle) C contained in S is a (p, q)-correspondence, defined by a single equation $G(u, v) = 0$.

To prove this, we can assume C to be irreducible. Take a point $a \in C$ and consider the cone T with vertex a that is the union of the lines joining a with an arbitrary point $m \in C$ (for $m = a$, take the tangent to C at a). A generator of this cone either intersects S at two points a and m, which lie on C, or is contained in S and must be one of the rectilinear generators G, G' of S that go through a. Thus the intersection $T \cdot S$ is of the form $C + jG + j'G'$. This intersection has an equation of the form $H(u, v) = 0$, where H has the same degree in u and v. Assuming, without loss of generality, that a is the point $u = 0$, $v = 0$, the polynomial $H(u, v)$ has $u^j v^{j'}$ as a factor because of the $jG + j'G'$ part (if $j \neq 0$, $H(0, v)$ is zero for all v, which implies one u factor, and so on). The quotient $G(u, v)$ gives the equation of C.

A (p,q)-correspondence is a curve (or cycle) of degree $p+q$.

Indeed, a plane tangent to S intersects it at $p+q$ points, p on one generator and q on the other.

A (p,q)-correspondence and a (p',q')-correspondence without common components have $qp'+pq'$ intersections.

To see this, let C and C' be the correspondences. After interchanging the factors in S, if necessary, we can assume $q \geq p$. Then, if G is an appropriately chosen generator for S, say $u=0$, $C+(q-p)G$ is a (q,q)-correspondence, and, by theorem A, a complete intersection $S \cdot T$, with $d^0(T)=q$. The points in common between $C+(q-p)G$ and C' are the common points of T and C', and there are $q(p'+q')$ of them. We remove from among them the points common to C' and $(q-p)G$, which form a cycle of degree $(q-p)q'$. There remain $q(p'+q')-(q-p)q'$, as we wished to show.

The degree-three curves contained in S are (apart from the degenerate $(3,0)$- and $(0,3)$-correspondences), the $(2,1)$- and $(1,2)$-correspondences already encountered; when irreducible, such twisted cubics are, as we have seen, unicursal.

There are irreducible degree-four curves on S apart from biquadratics, namely the $(1,3)$- and $(3,1)$-correspondences. These are unicursal curves because the equation of a $(1,3)$-correspondence gives u as the quotient of two degree-three polynomials in v. We call them *monoquadratics* because they are contained in only one quadric, S (otherwise they'd be complete intersections, which, as we know, are $(2,2)$-correspondences).

Notice that every curve of degree 4 is contained in a quadric. Indeed, the equation of a quadric has ten coefficients. The condition that this quadric passes through nine points of C gives a system of nine homogeneous equations in these ten coefficients, which has a non-trivial solution. The corresponding quadric has at least $9 > 2 \times 4$ points in common with C, so it contains C if C is irreducible. If C is a plane curve, the quadric will contain its plane.

The same result is clearly true for irreducible curves of degree three.

An irreducible biquadratic B is unicursal if and only if it has a double point.

Take a simple point $a \in B$ and project B from a onto a plane P. A plane going through a has three other intersections with B , so every line of P intersects the projection B' of B in three points; thus B' is a plane cubic. Its multiple points are either:

- the projections of the multiple points of B; or
- the points b of B' that are projections of several (distinct or not) points of B.

In the second case, D_{ab} would have these points, plus a, in common with B, hence also in common with S; then D_{ab} would be a generator of S, but such a generator intersects B at two points only. Thus this case is excluded, and we see that B' is unicursal if and only if B has a double point (theorem 39 in section 2.6). Since the coordinates of a point m of B are rational functions of its projection

m' on B' (for m is the point where $D_{am'}$ intersects again the quadric S), and conversely, B is unicursal if and only if B' is.

Composition of correspondences

Let C be a correspondence between two sets D and D', and C' a correspondence between D' and D''. The set of pairs $(m, m'') \in D \times D''$ such that there exists $m' \in D'$ for which $(m, m') \in C$ and $(m', m'') \in C'$ is a correspondence between D and D'', which we denote by $C' \circ C$ and call the *composition* of C and C'. In terms of the triple product $D \times D' \times D''$, this composition is expressed by

$$C' \circ C = \mathrm{proj}_{D \times D''}\big((C \times D'') \cap (D \times C')\big).$$

In algebraic geometry, we replace the point-set intersection by the *intersection product* $(C \times D'') \cdot (D \times C')$, and we give appropriate weights, called *projection indices*, to the projections.

For example, if E is a (p, q)-correspondence between D and D', its projection index onto D is q and we write $\mathrm{proj}_D(E) = qD$; this is because q points of E project onto one of D.

If C is a (p, q)-correspondence and C' a (p', q')-correspondence, C generally associates q points $m'_i \in D'$ to one point $m \in D$, and C' associates to each of the q points q' points $m''_{ij} \in D''$. Thus there are qq' points in D'' for each point in D, and the composition $C' \circ C$ is a (pp', qq')-correspondence. In particular, the composition of two $(2, 2)$-correspondences is a $(4, 4)$-correspondence.

When $D = D' = D''$ and B is a symmetric (n, n)-correspondence, its composition with itself, $B \circ B$, is reducible. Indeed, for each of the n points m'_i that correspond to a point $m \in D$, we have m among the n points m''_{ij} that correspond to m'_i. Thus m occurs at least n times among the m''_{ij}. If I denotes the identity, $B \circ B$ is of the form $nI + B'$, where B' is an $(n^2 - n, n^2 - n)$-correspondence. For $n = 2$, we have $B \circ B = 2I + B'$, where B' is a symmetric $(2, 2)$-correspondence. This explains why, in the Poncelet construction preceding theorem H, we encountered only $(2, 2)$-correspondences.

If D is a variety defined on a finite field $\mathbf{F}_q$, an important correspondence is the one that associates to a point of D with coordinates (x_i) the point with coordinates (x_i^q) (which is in D because the q-th power map is an automorphism of every algebraically closed field containing $\mathbf{F}_q$). We call that the *Frobenius correspondence*, and we denote it by F. The fixed points of its iterate F^n are the points of D that are rational over $\mathbf{P}_{q^n}$, because this field is the set of roots of $x^{q^n} - x = 0$. The estimation of the number of these fixed points is an important arithmetic problem; it was solved by A. Weil, then by P. Deligne, in their works on the Riemann hypothesis for fields of algebraic functions. The Frobenius correspondence F is a $(q, 1)$-correspondence, and F^n is an $(q^n, 1)$-correspondence. Since an (i, j)-correspondence over a projective line D has $i + j$ fixed points, namely, its intersections with the plane of the conic I that is the graph of the identity map on the Segre quadric, F^n has $1 + q^n$ fixed points in this case, which we knew anyway because that is the number of (rational) points of the projective line over $\mathbf{F}_{q^n}$.

Bibliography

Cited in the text

[Ar] Emil Artin, *Geometric Algebra*, Interscience, New York, 1957. Cited on page 34.

[Bo1] Nicolas Bourbaki, *Commutative Algebra*, Addison-Wesley, Reading, Mass. Cited on page 87.

[Bo2] Nicolas Bourbaki, *Espaces Vectoriels Topologiques*, Hermann, Paris, 1953. Cited on page 87.

[Bo3] Nicolas Bourbaki, *General Topology*, Addison-Wesley, Reading, Mass., 1966. Cited on page 87.

[Ha] M. Hall, *The Theory of Groups*, Macmillan, New York, 1959. Cited on page 33.

[Pi] Gunter Pickert (see below). Cited on page 33.

On projective geometry proper

Reinhold Baer, *Linear Algebra and Projective Geometry*, Academic Press, New York, 1952. An elementary exposition, from the algebraic point of view.

Abraham Seidenberg, *Lectures in Projective Geometry*, Van Nostrand, Princeton, 1962. Starts from high-school geometry, derives from that an axiomatization of the projective plane, then introduces coordinates.

Beniamino Segre, *Lectures on Modern Geometry*, Cremonese, Roma, 1961. A very complete exposition, at once algebraic and axiomatic. Starts with some preliminaries about rings and fields, and includes an axiomatic presentation of conics and quadrics, over general (non-commutative) fields. The appendix, by L. Lombardo-Radice, discusses non-desarguesian fields.

Robin Hartshorne, *Foundations of Projective Geometry*, Benjamin, New York, 1967. A brief and elegant treatment of the basics, from both the algebraic and the axiomatic points of view.

Gunter Pickert, *Projektive Ebene*, second edition, Springer-Verlag, Berlin, 1975. An axiomatic study of desarguesian and non-desarguesian projective planes.

Oldies but goodies

Elie Cartan, *Leçons sur la géométrie projective complexe*, Gauthier-Villars, Paris, 1931. Discusses the riemannian geometry of a number of spaces derived from complex projective geometry in dimension one and higher, in particular, the space of involutions and anti-involutions.

Ernest Duporcq, *Premiers principes de géométrie moderne*, third edition, Gauthier-Villars, Paris, 1949. Written at the turn of the century, this book starts off with a suddenness that would make shudder some of today's punctilious algebraists. Once this obstacle is overcome, one finds in it a wealth of beautiful results on conics, quadrics, twisted cubics, biquadratics, quadratic transformations, correspondences, etc.

More on algebraic geometry

Jean Dieudonné, *Cours de géométrie algébrique*, in two volumes, Presses Universitaires de France, Paris, 1974. The first volume gives a historical overview of the development of algebraic geometry; it is available in English as the *History of Algebraic Geometry* (Wadsworth, Belmont, Ca., 1985). The rest of the work is a presentation of the subject (including proofs) from one modern point of view.

Abraham Seidenberg, *Elements of the Theory of Algebraic Curves*, Addison-Wesley, Reading, Mass., 1968. Requires only a modest amount of algebraic knowledge; favors a concrete approach, devoting quite a bit of space to plane curves.

William Fulton, *Algebraic Curves*, Benjamin, New York, 1969. The necessary algebraic fundamentals are treated quickly and effectively. The tone is more abstract than in the previous book; the exposition leads up to the theorems of Riemann–Roch and Bezout, and to the resolution of curve singularities.

Alain Chenciner, *Courbes algébriques planes*, Publ. Math. Univ. Paris VII, 1979. Emphasizes the local study of curves, series expansions and the topology of curves and their singularities over the complex numbers.

Igor R. Shafarevic, *Basic Algebraic Geometry*, Springer-Verlag, New York, 1974. A fairly concrete treatment of contemporary algebraic geometry (sheaves, schemas, cohomology).

Robin Hartshorne, *Algebraic Geometry*, Springer-Verlag, New York, 1977. Like the preceding one, but more abstract.

Index of Symbols and Notations

Non-alphabetical symbols are listed first, grouped according to type. Greek letters are entered as if spelled out.

Index

Italics indicate a definition, or the statement of a result. Words used throughout the book are only indexed where defined or first used.

Undergraduate Texts in Mathematics